PROMOTING EMPLOYEE HEALTH

INDUSTRIAL RELATIONS IN PRACTICE
General Editor: Jim Matthewman

Industrial Relations in Practice is a new series intended for personnel managers, union negotiators, employees, welfare advisers and lawyers. With an emphasis on current practice in leading British organisations and trade unions, the series takes an overall independent stance, with titles aimed at both sides of industry. The various authors, who have been selected from management, independent research groups and labour organisations, address themselves to topics of immediate and practical concern to the work-force of today and those responsible for its management.

Edward Benson
A GUIDE TO REDUNDANCY LAW
THE LAW OF INDUSTRIAL CONFLICT

Gary Bowker
DISCRIMINATION AT WORK

Alastair Evans and Stephen Palmer
NEGOTIATING SHORTER WORKING HOURS

Philip James
UNDERSTANDING SHOP STEWARDS: Their Role and Legal Rights

Susan Johnstone and James Hillage
CONTROLLING INDUSTRIAL ACTION

David J. Murray Bruce
PROMOTING EMPLOYEE HEALTH

Susan M. Shortland
MANAGING RELOCATION

Peter Wickens
THE ROAD TO NISSAN: Flexibility, Quality, Teamwork

Series Standing Order

If you would like to receive future titles in this series as they are published, you can make use of our standing order facility. To place a standing order please contact your bookseller or, in case of difficulty, write to us at the address below with your name and address and the name of the series. Please state with which title you wish to begin your standing order. (If you live outside the UK we may not have the rights for your area, in which case we will forward your order to the publisher concerned.)

Standing Order Service, Macmillan Distribution Ltd, Houndmills, Basingstoke, Hampshire, RG21 2XS, England.

Promoting Employee Health

David J. Murray Bruce
Group Senior Medical Officer
National Westminster Bank plc

Foreword by Richard Schilling
Emeritus Professor of Occupational Health,
University of London

MACMILLAN

First published 1990

Published by
THE MACMILLAN PRESS LTD
Houndmills, Basingstoke, Hampshire RG21 2XS
and London
Companies and representatives
throughout the world

Printed in Hong Kong

British Library Cataloguing in Publication Data
Murray Bruce, David J. *1938–*
Promoting employee health
1. Great Britain. Industries. Personnel. Health.
I. Title. II. Series.
613.6'2
ISBN 0–333–42722–X

To Suzanne, Richard, Alexandra,
Kathryn and Sarah

Contents

Contents

List of Tables

Foreword

There is a growing awareness of the importance of occupational health in many countries. The United Kingdom has been one of the main protagonists through its Health and Safety at Work etc. Act of 1974. This Act covers virtually the whole of the British working population and has imposed additional responsibilities on both employers and employees, with much more stringent penalties for failure to comply with the law.

Although office workers had some protection under the Offices, Shops and Railway Premises Act of 1963, the Act of 1974 put new general duties on employers to provide and maintain a safe place of work and to consult with their employees. In Britain, office workers comprise some eleven million people out of a total work force of twenty-five million. While their exposure to specific occupational hazards of injury and disease is much less than factory and mine workers, their work and the conditions under which it is carried out, have considerable influence on health; and conversely the health of the individual has an effect on attendance and quality of work.

Office management needs to know the role that an occupational health service can play in helping to comply with the law and to achieve higher standards of health and safety. Dr Murray Bruce, who has first-hand experience in this field, has written one of the very few books devoted to office workers. He has covered a wide range of topics and described in non technical language the more common health problems which arise, such as heart disease, nervous breakdown, alcoholism and those affecting female staff. He shows how these problems can be dealt with. There are useful chapters on the organisation of a service staffed by an occupational health nurse; first aid requirements; accident prevention in the office, and the health of the office worker abroad.

This is a book which will be useful to firms who either have, or contemplate having, an occupational health service. It contains information which will be valuable to persons working in this field, particularly nurses. It will be especially helpful to the personnel staff of the many firms which are too small to have their own service.

RICHARD SCHILLING
Emeritus Professor of Occupational Health
University of London

Acknowledgements

I am grateful to National Westminster Bank plc for allowing me to write this book.

My thanks are also due to Professor Richard Schilling for his encouragement and support both for this book and throughout my career, and to those people, both inside and outside NAT WEST, who have helped me to produce this book.

I greatly appreciate the expertise and care taken by the editorial staff of Macmillan at all stages of production.

The views and recommendations in this book are my own and not those of National Westminster Bank plc.

DAVID J. MURRAY BRUCE

Introduction

GENERAL BACKGROUND

Up to World War II occupational health measures in this country were principally focused on high-risk occupations such as mining and in 'heavy' industry. Many of the hazards then no longer exist now and those that still do are closely monitored, with employees under health surveillance.

The health of office workers was largely taken for granted. The office environment is in general free from the hazards of the factory, but it can cause or contribute to ill health and accidents, in both obvious and more subtle ways. A number of Acts, in particular the Offices, Shops and Railway Premises Act of 1963 (OSRP) and the Health and Safety at Work etc. Act 1974 (HASWA), legislated to remove risks and set health and safety standards. Several service industry companies, employing many clerical staff, developed their own occupational health service, including National Westminster Bank.

The developed world has seen a technological revolution, a new industrial revolution. There is increasing automation and microchip computerisation, using the silicon chip and its successors, theoretically capable of ten thousand million calculations a second! Over half of the UK work force is employed in offices and this proportion is likely to increase, although total numbers may fall.

People work better if they are content. There is, for instance, a danger that morale will fall as a company grows, if an increase in size leads to anonymity with the loss of a sense of belonging. This was borne out by a Consumer Association survey of 24 000 members which showed that those who worked for small organisations were, in general, more satisfied with their jobs than those working for large organisations.

Factors contributing to job satisfaction also include the use made of individual abilities, interest in the work done, and being able to choose to some extent what work to do. It is particularly relevant to this book that the survey shows secretaries and clerical workers to be less satisfied with their jobs than people in other occupations.

* * *

1

There are two major differences in the philosophies adopted by companies towards occupational health for their employees. The first fulfils legal requirements and prevents and treats accidents and hazards encountered by employees at the work place. This is largely the aim of the Health and Safety Executive (HSE). The second philosophy includes the first, but in addition, considers the total health and well-being of the work force.

The discussion document on occupational health services – *The Way Ahead* – includes a survey of occupational health services in over 300 companies.

The survey showed that in round figures 30 per cent of manufacturing industry had some form of occupational health service compared with 11 per cent of nonmanufacturing industry. 85 per cent of the companies with 34 per cent of the work force had no occupational health service other than first aiders; 5½ per cent, employing 52 per cent of the work force, had either medical or nursing staff and only 2 per cent, employing 31 per cent of the work force, had medical and nursing staff.

Companies with less than 250 staff usually had no medical or nursing cover, or just a doctor on call. Exceptions were those companies which belong to group schemes. Such a scheme is usually based on an industrial estate. There is a central health centre for consultation, with medical and nursing staff visiting the companies in the scheme regularly as well as in emergencies. These companies pay an annual subscription per employee.

The objectives of an occupational health service have been defined by a Joint International Labour Office/World Health Organisation Committee as 'the promotion and maintenance of the highest degree of physical, mental and social well-being of workers in all occupations; the prevention amongst workers of departure from health caused by their working conditions; the protection of workers in their employment from risks resulting from factors adverse to health; the placing and maintenance of the worker in an occupational environment adapted to his physical and psychological condition.'

In this country, occupational health care dates from the early part of the nineteenth century, following awareness and concern about the harmful effects of working conditions on the health of those employed, especially children in the mines and in the mills and factories built in the wake of the Industrial Revolution. Although some disfiguring and disabling disorders were incorrectly associated with occupation, it is certain that many workers endured appalling working conditions. Also the living conditions for industrial workers and their families were often squalid, 'gaunt barracks'. Large numbers of people moved from unemployment in the country to work in industrial towns, often renting jerry-built, mean, cramped houses, cheaply and hastily erected, crowded together and having inadequate sanitation.

There were exceptions, the best known being Robert Owen's New Lanark and Manchester Mills, established in the early 1800s. Another mill owner, Sir Robert Peel the elder, had earlier championed the cause of factory children; he was responsible for introducing the Health and Morals of Apprentices Act 1802. In 1802, Charles Thackrah, a Leeds physician, published a book on industrial diseases 'to excite the public attention to the subject'. A series of parliamentary Factory Acts were passed, applying to mills and textile factories and later extended to much of industry. The Act of 1833 replaced voluntary inspection of factories by visitors with the Factory Inspectorate. At first, four inspectors were appointed to administer and enforce the new laws. These included regulations for employment of children: no more than twelve hours work a day (later cut to 10 hours), no night-work under eighteen and a minimum age for employment of nine; factory schools were also to be provided.

Each child was required to have a medical certificate of apparent age to replace the certificate which had been supplied by the parent. In the Factory Act of 1844, Factory Inspectors were empowered to appoint Certifying Surgeons to ensure the certificate of age was accurate and unbiased. Later these surgeons, at risk of being made redundant by national birth certification, certified a young person's fitness to work and also investigated accidents at work and occupation-related diseases. Thus developed the special involvement of doctors with people at work.

In 1898, Thomas Legge was appointed the first Medical Inspector of Factories and in the next thirty years he contributed much to occupational medicine. Many of his wise aphorisms remain as applicable today as when he wrote them.

The Certifying Surgeons later became Appointed Factory Doctors (AFD); these comprised some 1800 doctors, mostly part-time General Practitioners (GPs), who were responsible to the Medical Inspectors of Factories and involved with statutory medicals and the investigation of occupational disease and accidents. In 1972, the AFD service ceased and the Employment Medical Advisory Service was created.

* * *

In most European countries the emphasis in occupational health services is on preventive health measures: a pre-employment medical examination to ensure the applicant is medically fit for a particular job, periodic screening of 'at risk' categories of workers, and checks of the working environment. Treatment is confined to the emergency management of illness presenting itself for the first time at work, and first aid treatment of accidents.

However, in France, where there is no national health service, factory doctors contribute to overall health care. Since 1946 the services of a full-time doctor have been required for every factory or other work place where over 3000 people are employed; a part-time doctor is required where there are less than 3000. Group health care for adjacent small companies is encouraged with administrative assistance from the Ministry of Labour and Manpower.

In Holland, all work places with more than 750 employees are required to provide adequate industrial medical care; both the accommodation and the qualifications of those giving this care must meet certain requirements.

Similarly in West Germany, since the passing of the Work Safety Act in 1973, at any establishment with more than fifty employees, the employer is required to make safety specialists available, including a doctor and an engineer.

Since 1965 employers in Belgium have had a statutory duty to provide safety experts, including a doctor with specialist training. Work places with more than 2500 employees require more than one doctor; those with less than 2500 staff either employ their own doctor or belong to a group scheme.

In China, every factory and work place with over 2000 workers has both 'barefoot doctors', people who have shown an interest in health and are given up to six months of instruction, and a small clinic or hospital. This is run by one or more qualified doctors and has an operating theatre, pharmacy and modern methods of investigation.

The worker pays a small charge for these services. Preventive health is a major role of the 'barefoot doctor', who teaches such subjects as hygiene, the benefits of exercise and the dangers of smoking. Additionally, they diagnose and treat minor illness with traditional and modern drugs and also immunise and advise on family planning.

THE HEALTH AND SAFETY AT WORK ETC. ACT 1974

Before 1974 the laws on safety and health at work, contained in multiple statutes supported by hundreds of regulations, appeared haphazard and confusing, intricate in detail, difficult to amend and keep up to date, and enforced by no less than five government departments and seven inspectorates.

A new approach was needed to improve this and encourage participation by employees as well as employers, to get away from the concept that safety and health at work is a matter for regulation by outside agencies and 'is not my concern'. It was against this background that the Health and Safety at Work etc. Act 1974 (HASWA) was evolved.

The general purpose of HASWA is to achieve the following objectives:

(1) so far as reasonably practicable, to secure the health, safety and welfare of persons at work.
(2) so far as reasonably practicable, to protect persons other than those at work against risks to their health or safety from the activities of persons at work.

It is based on the 1972 report of the Committee of Enquiry on Safety and Health at Work chaired by Lord Robens.

The Act was introduced in March 1974, received Royal Assent four months later and the first provisions came into force on 1 April 1975.

The purpose of this 'Enabling' Act is to provide a legislative framework to promote, stimulate and encourage high standards of health and safety at work with effective safety organisation and performance, as well as to promote safety awareness by both employer and employee. There is the requirement of a general duty of care by everybody at work.

With this Act, the development of policies in the health and safety field passed from Government departments to the newly formed Health and Safety Commission (HSC). This consists of a Chairman and six to nine other people appointed by the Secretary of State for Employment to represent the Confederation of British Industry (CBI), the Trades Union Congress (TUC) and local authorities. The Commission has produced a series of explanatory booklets about its functions and about HASWA.

The Commission appoints, with the approval of the Secretary of

State, the Director of the Health and Safety Executive (HSE). The HSE puts into practice the policies of the Commission, enforces legal requirements (together with local authorities) and provides an advisory service to both sides of industry, called the Medical Services Division (see the section under this heading later in this Introduction).

The HSE was formed in 1975 and is responsible for the application, control and enforcement of HASWA. The HSE employs over 3000 staff, relocated from London to Bootle, and has nine divisions – policy, specialist services and operating ones. The latter consist of the Medical Services Division, the Research and Laboratory Services Division, a number of Inspectorates and the Technology and Air Pollution Division.

HASWA consists of four parts, of which Parts I and II are particularly relevant to this book. Part I relates to health, safety and welfare in relation to work, and Part II to the Medical Services Division. Part III amends the law relating to building regulations and Part IV contains general and miscellaneous provisions.

HASWA adds to, and does not supersede, existing health and safety legislation, in particular the Factories Act of 1961 and the Offices, Shops and Railway Premises (OSRP) Act 1963, both of which, in amended form, remain in force.

Unlike previous legislation, HASWA legislates for people and their activities rather than premises and processes. Under previous Acts there were many exclusions, totalling some five million people, including those employed in education and parts of the transport industry and leisure activities. Now, with the exception of domestic servants in private households, all persons at work are covered: employer, employee and the self-employed. In addition HASWA covers the health and safety of members of the general public who may be affected by work activities.

The acquisition, keeping and use of dangerous substances is included, as is control over the emission into the air of offensive or harmful substances.

Under HASWA, the employer of five or more of staff is legally required to prepare and publish a comprehensive health and safety policy.

OTHER RELEVANT ACTS

Every employee has the statutory right to work in safety. The employer has a common-law duty to the employees to provide, so far as is reasonably practicable, safe plant and equipment, that is the employer must provide safe premises and places of work, safe systems of work, adequate instruction, warnings of hazards, and competent colleagues. These requirements are enforced by a number of Acts apart from HASWA.

The Employer's Liability (Defective Equipment) Act makes the employer liable for faults in any equipment used. The Act has the effect of transferring common-law responsibility for negligence arising from faulty equipment from the importer, supplier or manufacturer to the employer. Employers have, in turn, a common-law right to sue the importer, supplier or manufacturer for any damages which are awarded against them under the Act. They can only do this, of course, if the importer, supplier or manufacturer is accessible and still trading. Employers must, and no doubt would anyway, ensure regular checking and maintenance of all equipment by suitably qualified persons. Repairs to equipment found to be faulty must be carried out promptly and the equipment not used until repaired. Examples are all types of electrical equipment, checking particularly the wiring and insulation, mechanical faults in typewriters, and faulty chairs and stepladders. Anyone can change, say, a light bulb, but if there is any doubt the advice must be 'leave it to the expert'.

Under the Employer's Liability (Compulsory Insurance) Act, employers must be insured against liability for injury or disease sustained by employees in the course of their work. *The certificate of insurance must be prominently displayed where all staff can see it.* There is a penalty for not doing this, as there is, on a daily basis, for an employer who is not so insured.

The Fire Precautions Act applies to the owner of the premises, rather than the employer who is the occupier and not the owner of the premises. However, it is a criminal offence for the occupier as well as the owner not to provide adequate means of escape for use in the event of fire.

Appointed under Section 18 of HASWA, Inspectors employed by the HSE and other enforcing authorities, to check standards of safety, have greater powers than before HASWA. They have the authority, when there is contravention of either this or other existing Acts, regulations or supplementary approved Codes of Practice, to serve an *Enforcement Notice*. This may be a *Prohibition Notice*, *immediate* or *deferred*, to stop an activity giving risk of serious personal injury (the term originally used in the Factories Act 1961 was 'serious bodily injury'; this was regarded as too restrictive since it did not embrace mental injury), or an *Improvement Notice* for remedy within a specified time of a fault which contravenes relevant statutory provisions. In addition, or instead of issuing either of these notices, an Inspector may prosecute in a Magistrate Court or the Crown Court with a fine on summary conviction. There is no ceiling to the fine for certain specific offences. The court can also make an order for the cause of the offence to be remedied. Failure to comply with Improvement or Prohibition Notices renders the person on whom the notice is served liable to prosecution and, in the case of a Prohibition Order, to imprisonment.

An appeal to an Industrial Tribunal against the notice, in full or in part, may be lodged within twenty-one days of service. To appeal against an Improvement Notice in order to gain time (the Notice being suspended until the appeal is heard) will be apparent to the Tribunal and dealt with accordingly. A Prohibition Notice will apply, irrespective of Notice of Appeal, once the time and date of its effect is reached.

The HSE statistics for 1987/88 show that at 11 139, 619 more enforcement notices were served than in the previous year, made up of 6623 improvement notices, 234 deferred prohibition notices and 4282 immediate prohibition notices. There had been a blitz on construction sites.

THE ENFORCEMENT OFFICER

The following is a summary of the powers of an Enforcement Officer, that is, a Factory Inspector of the HSE or an Environmental Health Officer.

Under HASWA an Enforcement Officer may enter any work premises, with a police officer if expecting obstruction, at any reasonable time to carry out, with the premises remaining undisturbed, such examinations and investigations as he considers necessary. These may include measurements, photographs, recordings and the taking of samples from articles, substances and the atmosphere, giving a portion of each to a responsible person representing the employer if it is practicable to do so, with particulars of the sample taken. He or she can:

have any article or substance considered dangerous dismantled or tested and can take possession of any such article or substance for examination or to ensure it will not be tampered with and will be available as evidence.

require anyone to provide facilities and assistance in relation to matters within that person's control or responsibility which will enable the inspector to exercise his or her powers.

require anyone he or she believes can give relevant information to answer questions and sign a declaration of the truth of the answers. The person giving the statement may nominate someone to be present and the inspector has discretion to allow anyone he or she thinks fit to be present. Any answers or statements given are inadmissible in evidence against the person who has given them and also against that person's husband or wife.

require production of, and may copy, any book or document required to be kept under HASWA or other relevant statutory provisions.

serve an *Improvement Notice*, giving the reasons and a time limit, on any person or body corporate contravening HASWA or other relevant statutory provisions. This puts the employer on notice to remedy the situation within a reasonable time.

serve a *Prohibition Notice*, again giving the reason, against any activity which in his opinion involves risks of serious personal injury. With an *Immediate Prohibition Notice* this activity must cease on receipt of the notice. A *Deferred Prohibition Notice* allows time for modification.

THE SAFETY POLICY

The policy statement is normally the first thing the visiting Enforcement Officer, from either HSE or the local authority, will ask to see when carrying out an inspection of a work place where five or more are employed. If the policy document is inadequate, the visit does not get off to a very good start.

Failure to provide a statement of general policy with respect to the health and safety at work of employees and the organisation and arrangements for this, is a contravention of the statutory requirements. The Enforcement Officer will make the first major assessment as to the degree of compliance with the requirements imposed on an employer by HASWA, by examining the statement.

The Enforcement Officer is usually prepared to advise, when

requested, those employers who have difficulty in understanding their obligations under HASWA or who need assistance in drawing up the safety policy.

The safety policy is a straightforward document which should cover three areas:

(1) The general statement – a short declaration of personal intent, written on headed paper and signed by the employer, to provide, with the employees' support, safe and healthy working conditions.

(2) The organisation for carrying out what is in the statement – the detail needed includes the names or appointments of those responsible, that is, safety officers, maintenance engineers, managers and supervisors. Also, the procedure for joint consultation must include the names of appointed safety representatives and the constitution of the joint safety committee.

(3) Arrangements for ensuring health and safety at work – this will be the longest part of the policy as it must cover the whole range of activities at work and include the following:
   (a) procedure for reporting and investigating accidents and 'near misses' at work;
   (b) arrangements for first aid treatment of accidents at work;
   (c) emergency procedures in the event of fire or explosion;
   (d) provision of protective clothing and equipment;
   (e) identification and means of dealing with hazards specific to the work place, for example, guards for machinery;
   (f) identification and means of dealing with common hazards, for example fire prevention and the use of ladders;
   (g) systems for regular inspection and maintenance of equipment;
   (h) safe systems and methods at work;
   (i) arrangements for advising employees about health and safety, both in general and specific to their particular work, especially when starting that work for the first time;
   (j) health and safety inspections.

Only an outline need be given in the policy document where there are separate detailed rules and procedures.

The statement must be drawn to the attention of all employees either by issuing a copy to each individual or posting a copy on notice boards in a prominent position.

The statement should be reviewed regularly and if updating or revision is necessary, these changes must be brought to the attention of each employee.

A health and safety training programme for all levels of employee should also be part of the safety policy; not only the specific on-the-job training referred to earlier, but integrated with this, more general health and safety instruction.

In the event of an accident at work, what is essential in order to prove a claim is that any accident, no matter how seemingly minor, must be reported to the employer or his representative as soon as possible by the employee or his representative. Of course, an accident must have caused injury, but the injury need not cause incapacity to work temporarily or permanently.

Most, if not all, of these requirements were already standard practice in the majority of companies. The difference now is that there is legal enforcement to ensure safe practices and prevent short cuts.

One important development is the emphasis on the involvement of the employees, the need for them to be safety conscious, identifying and reporting potential hazards. That safety is not the sole responsibility of management is clearly stated.

Within the scope of HASWA are the various preventive measures to avoid endangering health, such as the wearing of protective clothing, ear defenders, etc.

To minimise risks from the outset and to have a 'built-in' safety awareness, designers and manufacturers are required to ensure that, in so far as they are responsible, and given proper use, their articles will not endanger health and safety.

HASWA also makes provision for safety representatives, to be appointed by independent recognised trade unions, to represent employees in consultation with the employer on health and safety.

From 1 October 1978, it has been a criminal offence for employers not to consult with union-appointed safety representatives. These representatives are entitled to reasonable time off work with pay to pursue their safety role, be this to attend training courses, committee meetings or for investigation of possible hazards or after an accident at work.

Two representatives may demand that a Safety Committee be set up. Its functions, apart from investigation of accidents and potential hazards, include investigation into employee complaints relating to health and safety at work.

Having carried out an investigation, the Committee will then make representations to the employer. A representative is also entitled to meet, and receive information from, Inspectors.

At one time it seemed there was greater interest in getting more money from the employer for working with hazards or in a dangerous environment, for example, 'dirty money'; now there is a better realisation that long-term health and safety are more important.

THE MEDICAL SERVICES DIVISION

The Medical Services Division – previously the Employment Medical Advisory Service (EMAS) – has policy staff in London. There are also branches for planning and resource management, operations, assessment of health hazards, nursing and first aid, pathology and research. Employment Medical Advisors (EMAs) and Employment Nursing Advisors (ENAs) are based in thirty-two offices in ten regions. The Medical Services Division advises the Inspectorates on occupational health; performs statutory examinations of persons employed on hazardous operations; carries out investigations and surveys; advises the HSE, employers, trade unions and individuals on the occupational aspects of poisonous substances; undertakes research; provides advice on provision of occupational, medical, nursing and first aid services, and provides advice on the medical aspects of rehabilitation and training. It also sets standards for exposure to any substance which might be hazardous to health.

The EMAs and ENAs, like Environmental Health Officers, have the right of entry into any work place at all reasonable times and can insist that employers allow employees, if willing, to be examined during working hours.

If an occupational health hazard is found, they will advise on its reduction or elimination, using where needed HSE expertise either from the Inspectorates or the laboratory and if necessary will advise an Enforcement Officer.

EMAs monitor the statutory examination of factory employees exposed to certain health hazards, such as lead. Apart from these, routine statutory medicals are no longer required on young people at work.

A film explaining the work of EMAs, using case histories from factories, farms and a distillery, and entitled *Health at Work*, is available for hire from HSE.

Part I

Company Health Facilities

1 The Company Doctor

MEDICAL QUALIFICATIONS

A few employers appointed company doctors in the middle of the nineteenth century, but the voluntary appointment of company doctors in any numbers by employers was not stimulated until the Workers' Compensation Act was passed in 1897. The early appointments were intended more to protect employers against fraudulent claims than to protect the worker against accident and illness. This concept has proved difficult to erase.

For those involved in discussions with general practitioners about staff or in the selection of a company medical adviser, the following background information may be useful.

Abbreviations of professional qualifications are written after a doctor's name. Medical and surgical degrees are paired. The first part of a university degree, MB or BM, is Bachelor of Medicine (MD, Doctor of Medicine is the only qualification in the USA, but is a postgraduate qualification in this country, awarded for a successful thesis).

The abbreviation for the surgical qualification depends on the University – BS is Bachelor of Surgery, B.Chir and Ch.B are the Latin equivalents.

Alternative, or additional, qualifications are MRCS – Member of the Royal College of Surgeons and LRCP – Licentiate of the Royal College of Physicians. The longer number of letters in these qualifications has no special significance. Qualification from the Society of Apothecaries is signified by LMSSA.

In addition, there are many postgraduate diplomas and degrees. For example:

FRCS   – Fellow of the Royal College of Surgeons
MRCP   – Member of the Royal College of Physicians
MRCGP – Member of the Royal College of General Practitioners
MRCOG– Member of the Royal College of Obstetricians and Gynaecologists.
DCH   – Diploma of Child Health

A doctor does not usually put more than two of these abbreviations after his or her name, however many have been obtained.

On qualification, a doctor is provisionally registered with the General Medical Council (GMC), and after a year in approved hospital posts achieves full registration and his or her name is then included, subject to an annual retention fee, on the General Medical Register. In certain circumstances a doctor's name may be suspended or removed from the register ('struck off') temporarily or permanently. Anyone may check with the GMC whether a doctor is on the Register.

The Medical Directory is an independent, annually updated, two volume publication that lists mainly UK medical practitioners. Entry is voluntary. The information given is supplied by the doctor and includes medical school, year of qualification, and a précis of medical experience and interests.

OCCUPATIONAL MEDICINE TRAINING

In the UK, around 800 doctors work full-time and 2000 part-time in occupational medicine, out of a total of 93 000 practising doctors. Half the working population has no medical cover at work.

Doctors who specialise in occupational medicine have a great interest in preventive medicine, health promotion and the detection of occupation-related diseases. A fringe benefit is 'regular' office hours of work, that is, 9 a.m. to 5 p.m. free from night and weekend calls. Whereas in the past, recruitment of doctors to run an occupational health department had been by personal recommendation, expertise and experience of occupational medicine are now expected. The Universities of London, Birmingham, Manchester, Edinburgh, Surrey and Dundee run courses in occupational medicine full-time, on day release or by distance learning to prepare candidates for Associateship of the Faculty of Occupational Medicine (AFOM), which has largely replaced the Diploma of Industrial Health (DIH). In addition, the TUC Centenary Institute of Occupational Health at the London School of Hygiene and Tropical Medicine runs a full-time course for one academic year or part-time for two years which leads to the Master of Science (MSc) degree in occupational medicine. Doctors who work in this field are usually members of the Society of Occupational Medicine, which holds regional meetings during the academic year and produces a quarterly journal.

In 1977, the Faculty of Occupational Medicine was created within the Royal College of Physicians. Initially, entry was by invitation or application from doctors experienced in the speciality, that is, with at least twelve years' practice of occupational medicine, with at least three years in charge of an occupational health unit. Postgraduate qualifications and publications were taken into account. Membership of the faculty is denoted by the letters MFOM: examination by way of a dissertation is now needed by an AFOM to achieve this. Eminent members are elected Fellows.

CAREER STRUCTURE

The British Medical Association (BMA) issues guidelines of career and salary structures, for doctors practising occupational medicine, which are regularly reviewed. The Faculty of Occupational Medicine has formalised and standardised the training necessary to achieve specialist status in occupational medicine.

Promotion prospects are limited and this is reflected in the BMA minimum salary guidelines. Part-time Medical Officers are paid a sessional rate, or an hourly rate if weekly commitment is less than one session of up to three and a half hours. The lowest grade, Occupational Physician (Senior Registrar), is for a doctor experienced in general medicine but inexperienced in occupational medicine, training under supervision for a career in occupational medicine. This is comparable to the hospital grade of Senior Registrar. Time is allocated for training in working hours, for example, a working day each week to attend a day-release occupational medicine course.

A training scheme that lasts four years and involving appointment to an approved training post is recommended by the Specialist Advisory Committee of the Joint Committee on Higher Medical Training (JCHMT). Working for several contrasting companies provides wider experience, but may not be practicable. This grade should not be held for more than a limited time, after which the doctor enters the *full* Occupational Physician grade. This is also the grade recommended for a doctor after specialised training or with relevant experience who is in charge of medical services for a small organisation or a section of a large organisation.

A Senior Occupational Physician is usually promoted from the Occupational Physician grade, is in charge of the medical services of a large organisation, and probably controls other doctors and therefore needs administrative skills. The most senior rank is a Chief or

Principal Occupational Physician, who is Director of the Medical Services of a major company.

The Doctor/Occupational Health Physician (OHP) responsible for the occupational health of a company should be responsible to, and report directly to, a member of the board of that company.

With few or no medical colleagues at their place of work, many OHPs are medically isolated, compared with those in hospital medicine or general practice, and to keep in touch with current medical practice, need the opportunity to attend daytime professional meetings and short full-time courses. Though occupational health care is largely preventive and treatment is limited to emergency measures, the company doctor may be sued for alleged error or neglect, and must be a member of a medical defence organisation such as the Medical Defence Union or the Medical Protection Society. For an annual fee, these organisations advise and act for the doctor and pay the damages awarded by a court if negligence is proved.

FUNCTIONS AND RESPONSIBILITIES

In their role of promoting the health of the staff in the Company, company doctors are primarily concerned with the effects of health on work and the effects of work on health. They need to have detailed knowledge of work processes and the working environment, and their possible effects on health.

Their duties can be summarised as follows:

(1) Advise management on:
    (a) the working environment in relation to health;
    (b) the occurrence and significance of hazards;
    (c) accident prevention;
    (d) statutory health requirements.
(2) Advise employees on health matters relating to their work.
(3) Examine applicants for specific jobs.
(4) Treat medical emergencies and injuries which occur at work.
(5) Examine and follow up employees on their return after sickness absence and advise on suitable work.
(6) Liaise with general practitioners, hospital consultants, Employment Medical Advisers and Disablement Resettlement Officers about the health of an employee, with the informed consent in writing of that employee, as per the Access to Medical Reports Act 1988.

(7) Examine periodically certain categories of staff because of their work, for example, drivers and staff travelling abroad for the company (and the expatriate's accompanying family before and optionally after a tour).

(8) Supervise the health of all employees at work, especially disabled persons and catering staff.

(9) Supervise hygiene of staff facilities, for example, catering units and dining rooms.

(10) Be in overall charge of nursing and first aid services.

(11) Advise the Safety Committee.

(12) Promote preventive medicine and health education.

Employers sometimes grumble that doctors side with their patients and issue medical certificates too readily; a company may appoint their own doctor as their way of dealing with this – 'set a thief to catch a thief'. The newly-appointed company doctor will correct them on this point and explain that what he or she can do is find out the medical position and that the employer has no right, without informed consent, to have reported to him any information given to the company doctor by employees or their own doctors. The Access to Medical Reports Act further formalises this. The view that, as the company doctor uses company headed paper, all records are the property of the company, is offensive and incorrect, as is the claim that, as patient and doctor are both employees of the company, the normal doctor-patient relationship does not exist. There will be occasions when the doctor is acting solely for the employer when this must be made clear to the employee. A doctor can be compelled to disclose medical records in a court of law following a compulsory, statutory medical examination, when the doctor must disclose the outcome.

An employee may try, perhaps subconsciously, to play one doctor off against another, or to obtain a second opinion because of dissatisfaction with their GP or specialist; the company doctor avoids getting drawn into this situation. The company doctor also avoids prescribing for a non-urgent matter, for an employee who merely finds it inconvenient to consult his or her GP, or who, having moved house and area some time before, has still not registered with a new GP.

Occupational health is not primarily a treatment service, but an employee who for some genuine reason has not been able to consult his or her GP may reasonably ask to see the company doctor; this is in order for the company doctor to give an opinion and emergency treatment and, if necessary, to refer the employee to the GP with a letter. The company doctor may be able to help both the employee and the GP by arranging a specialist appointment local to the work place, with the specialist's report going to the GP.

The management of a company which employs a doctor must understand and accept that the company doctor has a strict ethical code. Advice given by this doctor may or may not be acted on by management. The suspicion of the work force, that the company doctor, seen as part of the management function, is the 'employer's man', lingers on. The doctor is and must be seen to be impartial, not a tool of management. What staff say at medical interview is confidential, unless informed consent has been given to disclose to personnel management factors relevant to that person's work. The company doctor will give an opinion to management on the medical suitability of an employee or applicant for a particular job, but without informed consent, will reveal no more. When an applicant's health questionnaire has been checked and where necessary further medical history obtained, or examination arranged, all that need usually be disclosed is whether that person is medically suitable or unsuitable for the proposed job: the potential employer is obviously concerned about the likelihood of lengthy or repeated sickness absence. If a medical condition precludes the particular job applied for, but not other jobs in the organisation, then, with consent, it is reasonable for the doctor to discuss this with management. Each time the company doctor seeks a report about an applicant or employee from that person's GP or specialist, written dated consent must be sent with the enquiry.

Until 1989, signing a consent form to allow an approach by the doctor acting for an employer for medical details to a GP or specialist, was like signing a blank cheque, with the employee having no clear idea exactly what would be disclosed. The Access to Medical Reports Act 1988 gives the individual the right, if he or she so chooses, to be fully aware of what a Company (an employer or Insurance Company) is being told by that person's doctor in a report about their health. There is also the opportunity for the employee to have input into that report to correct what he or she sees as wrong or misleading information. What follows is written as it would be explained to a member of staff:

Since January 1989, as a result of a private member's bill, you have the right, if you want, to read a medical report written about you after that date for an employer or insurance company by a doctor who has been responsible for your clinical care. If you indicate on the consent form that you want to read the report, you have twenty-one days to do so from the day the report is asked for. The doctor is told in the enquiry letter and can see on the consent form whether you want access and you will be told at the same time the enquiry letter is sent. It is then up to you to arrange with your doctor to read the report once it has been written and before it is posted back. You may for a fee have a copy of the report. The options then open to you are given on the consent form (and also on the cover of a pre-employment health questionnaire) namely:

(a) confirm consent to the report being issued
(b) request amendment of the report before submission
(c) attach a statement to the report reflecting your views, if your practitioner will not agree to the amendment
(d) withdraw consent to the report being issued.

If your doctor denies you access to part of the report (the Act permits this in certain circumstances), you may see the remainder. If you are denied access to all or part of the report, you retain the right to withdraw consent to it being issued. [A doctor is likely to show the report to their patient whenever possible, but is not obliged to disclose anything that would be likely to cause serious harm to that person's physical or mental health, or to disclose information about someone else] You may still ask your doctor for access to the report during the next six months.

There is also talk of a similar Act to cover correspondence between doctors about their mutual patient, for example a referral letter from a GP to a consultant.

A doctor who has received an enquiry letter outlining the company's concern, together with a consent form including the following phrasing, 'I agree to your submitting a medical report to the company doctor so that advice can be given to the company. I wish/do not wish to have access to the report before submission', will clearly know why the report is needed. On receipt of the report the company doctor will disclose to personnel management only what is relevant to that person's work, adding his or her own opinion. What is above all needed by the employer from a report, along with the current position, is the prognosis.

To refer an employee to the Company doctor because of a poor sickness record is an option personnel management may take. A manager who is concerned about the health of an employee will also suggest to that person that he or she consults their GP. The company doctor needs to know why the company is concerned. The type of format shown on page 23 can be used by personnel for this purpose and also gives other relevant information about the employee. In some cases, before seeing an employee, the company doctor will seek details, with consent under the Access to Medical Reports Act, from the GP and/or specialist.

In summary all staff should realise that

(1) Once in possession of all the facts, the company doctor gives an independent medical opinion which personnel management takes into account when making their decisions.
(2) All doctors abide by a strict code of medical ethics and medical etiquette.
(3) What an employee says to a doctor is confidential and not divulged without consent.
(4) A doctor's first responsibility is the health of the patient.
(5) A company doctor has no disciplinary role.

The company doctor's report is sent to the personnel manager but the line manager or supervisor also needs to know when someone in their unit has a condition, such as diabetes or epilepsy, which could cause sudden illness at work.

USE OF A GENERAL PRACTITIONER

Personal recommendation is the best way to choose a GP and a meeting should be arranged with that doctor before asking to be added to his or her NHS list. Post Offices, public libraries and local offices of the DHSS and Community Health Council hold lists of local GPs. A GP can arrange investigation or referral to a specialist or in special circumstances arrange for the specialist to visit the patient's home.

Patients should where possible visit the GP's surgery but those too ill, or with other good reason, can ask the GP to visit them at home.

The GP issues medical statements required under various Acts for claiming benefits.

In some rural areas the GP dispenses drugs and medicines, otherwise treatment by prescription is dispensed by a pharmacist. A prescription charge is made except during pregnancy and for children under sixteen years of age, women aged sixty and over, men aged sixty-five and over, those suffering from certain specified conditions

FIRST REPORT TO OCCUPATIONAL HEALTH DEPARTMENT

SURNAME: FIRST NAME(S):
(MAIDEN NAME):

BRANCH/DEPARTMENT: STATUS:

HOME ADDRESS:

DATE OF BIRTH: DATE OF ENTRY:
                                        NORMAL RETIREMENT AGE:

DATE FIRST ABSENT   MEDICAL CERTIFICATE EXPIRY
ANTICIPATED RETURN TO   DATE:
    WORK:

DETAILS OF WORK PERFORMED:

DATE OF PREVIOUS REFERRAL:
REASON FOR REFERRAL:

DETAILS OF RELEVANT DOMESTIC & WORK RELATED FACTORS:

SICKNESS RECORD
COPY ENCLOSED: YES/NO

MEDICAL FORMS OF CONSENT   NAME & ADDRESS
    TO:–
(Telephone Number/Hospital
    Reference Number)

ASSESSMENT FROM PERSONNEL

Please provide a medical report on our above-named member of staff, to
include a prognosis for the future and recommendations regarding any
options which the Company should consider.

SIGNATURE:
TITLE: DATE:

and certain low-income groups. A person needing regular medication
by prescription can buy a season ticket.

A patient away from home on holiday or business in the UK may
consult a local GP as a temporary resident.

When a company has branches throughout the country, as with, for
instance, chain stores, supermarkets and banks, a centralised Occupational Health Service at Head Office may be criticised as being

unavailable to all staff. One answer is to arrange for local GPs to visit company units in their areas to see staff who have needed sick leave. For this, the doctor is paid a retainer, plus additional fees for pre-employment examinations, and so on.

An alternative is for the occupational health service to have a central base from which doctors and nurses regularly visit the branches. A disadvantage of this is the time spent travelling. Another alternative is a team of locally based Regional Nursing Advisers.

A large national company is likely to have full-time or part-time company doctors and/or nurses in each region, and an international company, in each country.

On large industrial estates there are successful schemes providing occupational health services (OHS) for a group of unrelated companies. This service is based on a central unit where certain treatments like physiotherapy are provided and doctors and nurses are available for consultation. These OHS staff also regularly visit the member companies. Running costs are covered by annual subscriptions, paid by each member company for each employee in the work force.

2 The Company Nurse

There is no legal requirement for a company to employ a nurse, no matter how many staff are employed. However, in the absence of any particular hazards, for instance, in an office setting, a rough guide is to employ a full-time nurse when there are more than one thousand staff in a building or local area.

NURSING QUALIFICATIONS

About 211 000 nurses work in the UK. The title 'nurse' denotes a man or woman who has successfully completed a theoretical and practical General Nursing Council (GNC)-approved syllabus of training in approved hospitals, passed the necessary examinations and been entered on the register. Many present nurses are SEN or SRN, but all now qualifying become RGNs (Registered General Nurse) with the UK Central Council for Nurses, Midwives and Health Visitors. An SEN will have trained for two years and an SRN for three years. SEN training was essentially practical, but has been discontinued. Many SRNs also became midwives (SCM) which involved a further year of training; this makes them well able to advise on aspects of women's health.

OCCUPATIONAL HEALTH NURSE TRAINING

There is no legal requirement that nurses working in industry should possess the Occupational Health Nursing Certificate (OHN Cert.). However, particularly where a nurse but no full-time doctor is employed, there is increasing awareness by employers of the value of OHN Cert. training. An OHN Cert. trained nurse will have specialised knowledge of the work place that a part-time doctor visiting, say, one half-day each week, may not have. Practical experience is no substitute for specialised training.

The Royal College of Nursing (RCN) Institute of Advanced Nursing Education is both the training and examining body for Occupational Health Nurses. A one year full-time or two year day-release course must be taken to enter the OHN Cert. examination; there is the additional hurdle of an interview and essay to prove suitability for and entry to the course in the first place. The courses are held at either the RCN or at one of the seventeen RCN/UKCC approved centres in the UK.

The syllabus includes: aims, organisation, types and functions of occupational health services, and the responsibility and relationships of the occupational health nurse in the work place and outside; the functions of the occupational health nurse, including supporting the first aiders and later the rehabilitative care of those injured or taken ill at work; the prevention of ill health and conservation of health; the principles of workplace monitoring and the prevention of hazards; the assessment and maintenance of mental and physical health in relation to work; counselling; the planning, implementation and administration of first aid in accordance with the needs of the employer; the keeping and use of records; elementary statistics; epidemiology; communication skills; supporting services; and legislation concerning the individual and the environment.

There is usually plenty of interest in full or part-time nursing posts in industry. Forty percent of nurses want to work part-time; most are excited by the challenge of occupational health; some nurses may once have been attracted because of the hours – no nights or weekends on duty; and a few by the prospect of a quiet, ordered, peaceful existence, sitting waiting for employees to visit the unit: in other words, a reactive service. Having 'surgery hours' avoids the need for this. In my experience and contrary to lay expectation, first aid takes up only a fraction of the nurses's time and in a comprehensive dynamic occupational health service the nurse has a much more positive and cost-effective preventive role.

In the NHS, the more senior a nurse becomes, the more administrative is the work. Some nurses miss being with patients and it is the knowledge that they will get back to this that may start an interest in occupational health. The RCN recommends scales of pay and the following career structure for nurses in industry.

CAREER STRUCTURE

Occupational Health Nurse

A nurse working in industry without formal OHN Cert. training. In these cases the RCN recommends that such training be arranged as soon as possible.

Occupational Health Nursing Officer (OHN)

A nurse working in industry either alone in a small unit or directly supervised by an Occupational Health Nurse of a higher grade.

Senior Occupational Health Nursing Officer

A nurse with the OHN Cert. working alone or without direct supervision who also has a managerial role and is responsible to senior management or to a Chief Occupational Health Nursing Officer.

Chief Occupational Health Nursing Officer

A managerial administrative post held by a nurse with the OHN Cert. who is experienced in OH nursing and has responsibility for nursing policies and control of nursing functions in a large company.

FUNCTIONS AND RESPONSIBILITIES

In the past, management and unions were somewhat wary of the company nurse, their needless concern being that a nurse, if allowed in the work place, will find 'nit-picking faults' and insist on their correction – 'the tail wagging the dog' – so causing resentment and generally 'stirring things up.'

That the nurse is not a 'first-aider' is now better appreciated with the realisation that the nurse is a highly-trained professional who will not have job satisfaction if the work is no more than first aid. The company nurse trained in occupational health is a professional ally to both worker and manager, is always aware of confidentiality, and has a trained eye to spot potential problems on site. Subject to the normal courtesies of giving notice of intended visits, he or she should have unrestricted access to every part of the office or factory. The OHN Cert. gives authority to the nurse in this field and qualifies her or him to comment and advise on a wide range of occupational health matters, as the following job description illustrates:

(1) Provide a comprehensive occupational health nursing service for the benefit of all staff.
(2) Provide, in liaison with the first aiders, nursing care for injury or illness which occurs at work.
(3) Provide follow-up treatment of injuries sustained at work and of other medical conditions which can conveniently and

effectively be treated at work, in liaison with the employee's GP or specialist.

(4) Give health education and advice to individuals and groups of employees on health matters generally, but particularly relating to hygiene and occupation.

(5) Be available for counselling, when employees can discuss personal or social problems.

(6) Keep adequate records which may cover:
  (a) all attendances for treatment or examination;
  (b) all certified absences;
  (c) accident reports;
  (d) records as required by HASWA and other statutes, regulations and codes of practice;
  (e) environmental and statistical surveys.

(7) Ensure that occupational health nursing services are available during normal working hours.

(8) Maintain liaison with GPs, consultants, social workers, and other occupational health nurses, hygienists and doctors.

(9) Co-operate with EMAs and ENAs and with Safety and Welfare Officers.

(10) Provide or arrange for the first-aid training of employees as required by the first-aid regulations of HASWA and for appointed persons in first aid.

(11) Management training in aspects of occupational health.

(12) Hygiene visits to staff catering units and toilets.

(13) Be involved with the health aspects of pest control and refuse disposal, in liaison with the Environmental Health Officer.

(14) Follow up employees who have been absent from work on sick leave, on their return to work.

(15) Advise management on pre-employment health screening.

(16) Arrange immunisation of staff travelling abroad for the company.

(17) Carry out routine health surveillance including investigations such as hearing and lung function tests.

(18) Participate in Health and Safety committee meetings.

(19) Investigate, advise and monitor, in conjunction with the Occupational Hygienist or Safety Officer, or both, where these specialists are also employed, health aspects of the working environment, consulting and advising line management as necessary.

A lay manager is not qualified to judge the quality of the nurse's professional work. If the nurse sees a lot of staff each day, this does not necessarily mean she or he is doing a good job. Better may be the nurse who sees relatively few each day and whose health promotion advice has reduced the incidence of minor illness and of accidents.

Some companies employ a full-time nurse and arrange for a weekly visit from a doctor (for example, a local GP who may have had no formal postgraduate training in occupational medicine). Where a company is of sufficient size, one or more doctors may be employed full-time: such a doctor is a member of staff, is fully committed to the company, and is eligible for all benefits such as a pension, paid annual leave and paid sick leave. The company nurse enjoys the same benefits.

All staff must appreciate that neither the company doctor nor the nurse have disciplinary roles. They will see staff about whom the employer is concerned because of their health, advising on the effects of their health on their work and their work on their health, for instance after long-term sickness absence or repeated short-term sickness absence, and will advise the employer whether health is affecting present ability and give an opinion of the prognosis, and future fitness to do the job. They will advise whether the job or the work station requires modification, such as a change of chair, office or hours of work.

They may quite properly treat and investigate accidents that occur at work, and commence treatment for any acute illness, major or minor, which starts during working hours. Treatment is only given for that day and further treatment is arranged by referral to hospital for an emergency, or to the GP. It is the GP who is responsible for the overall primary health care of the patients registered with him or her, not the company doctor or nurse, who are unlikely to be available outside office hours and at weekends. To continue to treat a non-urgent patient without reference to the GP is not only unethical, but may also create difficulties for the GP who will have had no information about the case if the patient later contacts him or her. However, continuation treatment may be given at work when convenient, but always after liaison with the GP.

A typical letter from a company nurse to a GP may read:

Dear Doctor,

. . . . (name of patient)
The above named patient of yours attended the occupational Health Unit today complaining of

The following treatment was given.

I have advised him/her to consult you.
Please let me know if I can be of further help.
Yours sincerely

We seek to help employees remain at work or return to work after illness or injury. Facilities are available at work for such treatment as injections, re-dressings and physiotherapy. When possible, alternative employment is found for persons fit for work, but temporarily unfit for their usual job.

As will be clear, the emphasis of a company nurse's work is on well-being and prevention, but he or she may be involved in treatment and have access to a standard range of drugs with prescription only medicines, agreed with the company doctor in writing, and reviewed each year.

3 The Occupational Health Unit

The ideal is for the rooms in the unit to be purpose-built and to involve the company doctor and nurse from the planning stage. Where existing rooms have to be adapted, some of the suggestions which follow may not be practical but should give ideas for a successful adaptation.

Ready identification is essential. The title given to the rooms is important. 'First Aid Room' would be appropriate if there is no doctor or nurse present, but when either or both is employed this description takes no account of the many other duties performed there. 'Sick Room' over-emphasises sickness rather than health promotion, negating the preventive medicine approach which is the core of occupational health. 'Surgery' is misleading – the only surgical procedures will be the dressing of injuries and the removal of stitches. 'Nurse's Room' is descriptive but could indicate a dental or children's nurse, nanny, or so on. Nursing Officer or Adviser is the title for a company nurse, or occupational health nursing officer (OHN) if OHN Cert. trained. Either of these terms makes an unwieldly room title. 'Health Services Unit' suggests health promotion, 'Occupational Health Unit' is an alternative. It is inappropriate to include 'medical' in the title unless a doctor is present. The title 'department' signifies a certain minimum size and may not correspond with company structure if there is but one nurse in two rooms.

The occupational health unit must be readily accessible and well advertised. The extension number could be on every telephone. Concern that this will encourage staff to visit without good reason in order to take a break from work is not borne out in practice. The nurse's attitude is all important and he or she should recognise and discourage this trend. The nurse is an adviser with decisions about staff movement taken by management. Every office notice board should list the location of the unit, the nurse's hours of work and telephone extension.

The unit must be readily accessible to ambulance staff and convenient to an exit for transport to hospital or home. Rooms in the unit require good natural light, privacy and adequate soundproofing between rooms in the unit, the unit and the corridor (preferably not a main corridor), and the unit and adjoining offices.

The unit may comprise a treatment room, rest room, office, waiting area and toilet but with increasing emphasis on prevention rather than treatment this can be scaled down. General requirements for the unit include pastel-coloured washable walls; washable cushion-tread non-slip vinyl flooring (with carpet in the office area only); coved floors and ceilings to avoid dust traps; vertical slat blinds for privacy (these collect less dust than venetian blinds); 13 amp power points at convenient heights (that is, at working level in the treatment room); doors wide enough to take a wheelchair; hot and cold water supply; mains drinking water; battery beacons in case of power failure; 'enter' and 'engaged' calling system for waiting area; clear notices showing the location of the unit; and cleaning equipment for the unit in adjacent corridor cupboard.

EQUIPMENT NEEDED

Treatment room

This room will be used by first aiders when the nurse is out. There should be a wall-mounted Anglepoise lamp, a disposable-towel holder, a paper-cup holder, a clock with sweep-second hand, a stocked first aid box, coat hooks and a soap dispenser. Also required are a stainless-steel sink unit with elbow taps and adjoining laminated work surfaces with rounded edges coved to the wall and with lockable kitchen-unit-type cupboards underneath and overhead. Add a Snellen eye-test chart, an electric antoclave for sterilising reusable equipment, a small refrigerator (for vaccines and blood samples), an examination chair with headrest, a waste-disposal unit, an examination couch, screens on castors, a mobile illuminated magnifying glass, a carrying-chair/stretcher, a weighing machine kg/lbs plus height attachment cms/ins, a stainless-steel pedal bin, a dressing trolley, stainless-steel scissors, splinter forceps, dressing forceps, a tongue depressor, stainless-steel dressing bowls, galley pots, kidney dishes, a stainless-steel jug, a stethoscope, an auroscope (for examining ears), a blood pressure machine, glass dressing jars with stainless-steel lids, splints for fractures, a tendon hammer, a tape measure and a dressing gown/disposable gowns.

There must be the means for the safe disposal of used disposable syringes, needles, dressings, and so on, with arrangements made for collection and incineration. If syringes and needles are not carefully disposed of, they may cause a needle stick injury or be retrieved and abused and transmit hepatitis or AIDS. The disposable needle and syringe are put straight into a special sealed container.

The storage cupboards should have hinged doors; glass is best avoided as are sliding doors: their tracks collect dust and are difficult to clean. The daily cleaning of the unit must be to a high standard: twice a day is best, at midday (morning is when the unit is usually busiest), and in the evening.

Office equipment

This should include a wall clock with sweep-second hand, an automatic kettle, a metal waste-paper basket, an internal telephone, a GPO telephone, a chair and a desk with locking drawers to hold (1) file cards for individual staff records and (2) an A4 filing drawer; a calendar, a controlled drugs cupboard (that is, a locked cupboard within a locked cupboard), free-standing bookshelves, a cork notice board, a locking hanging cupboard for the nurse's personal belongings, and a room thermometer. (The office can be used as a visiting doctor's consulting room.)

Waiting area

This should have an internal telephone, two chairs, a cork notice board, chemical and water fire extinguishers, an entrance buzzer and a pamphlet stand or rack.

Rest room

This should be equipped with two 2'6" or 3' divans with firm mattresses, with hardboard between mattress and base, screens for privacy, coat-hooks, two chairs, dimmer facilities for the lights and provision for total darkness when required.

Toilet

This should be suitable for access by wheelchair, with grab rails, toilet seat, handbasin, lever door handles and a lock that can be opened from the outside. Urine analysis equipment can be stored in the toilet.

4 First Aid Requirements

HEALTH AND SAFETY (FIRST AID) REGULATIONS 1981

These regulations, comprising an approved code of practice with guidance notes for first aid at work, came into force in 1982, replacing the former multiple, rigid, legislative requirements with a single set of more flexible regulations and a code of practice which, with very few exceptions, applies to all those at work, including for the first time hotel and school staff and the self-employed. (An anomaly is the absence of first aid cover for school children and students.) A code of practice is not legally binding, but is regarded by the Courts as relevant when there is a failure to comply with a provision of the regulations.

Adequate facilities and equipment and suitably trained personnel must be provided. Criteria for deciding the extent of first-aid cover include, apart from numbers of employees, the nature and hours of the work, the layout of the work place, the scatter of the work force and the distance from medical help. For offices the general rule is still as it was under the OSRP Act: one first-aider once there are 150 employees, but the ratio is now one first-aider *per* 150 employees. Most office settings will have fewer than 300 staff so legally need no more than one first-aider, but provision must be made to cover that person's lunch hour, annual leave, sick leave and so on, so in practice two first-aiders are needed. The numbers criterion is of course very artificial and it would be ideal for every work place to have someone who knows the essentials of first aid.

In general, a first aid room is needed when there are 400 or more employees, with additional first aid kits in any areas from where it would take more than approximately three minutes to reach the first-aid room.

In a work place where potentially there is a special hazard, from a toxic chemical for example, the first-aiders must know what to do if there is any acute adverse effect on health. Such first-aiders need additional training.

For employees working alone or in small groups in isolated locations, small, travelling first aid kits should be provided. It is useful for as many of the work force as possible to be trained in emergency first aid. In the guidance notes for first aid equipment, the recommendation for a pressure bandage, which acted as a tourniquet on a bleeding limb but which would cause gangrene if left in place too long, has thankfully been deleted. Where there is no nearby tap water, containers of sterile water solutions should be available; these are also used if a chemical is splashed in the eye. Litre containers of sterile saline are convenient; they have an expiry date to be checked and need clear instructions.

Clean garments should be provided for use by first-aiders in the first aid room. The first-aider should be a volunteer, not 'press-ganged' into the job; a cool, calm person, not put off by the sight of blood, and readily available if needed. Training courses are approved by the HSE and overseen by Employment Nursing Advisers, who spot-check to ensure that standards are maintained. An initial course lasts four full days. A refresher course, taken every three years before the existing certificate expires (overlook this and it's 'snakes and ladders', back to a four day beginners' course), lasts at least one day. Half this refresher day is taken up with the exam, so only half a day is left to practise resuscitation, revise existing knowledge and update on any changes. For this reason two days is a better length for this course, especially as, unless the first-aider belongs to the British Red Cross Society or the St. John Ambulance, there may have been few if any opportunities in an office environment to carry out any first aid other than perhaps treating a paper cut! First-aiders can soon lose confidence, especially in resuscitation skills. Specialised training relating to a special work-place hazard may be given by an outside agency or in-house at the work place if the employer has the means. Records must be kept of what has been taught to whom and when. When there is no requirement for a first-aider, the appointed person and their deputy, responsible for maintaining the first aid box and for making the call for an ambulance, should ideally have had a short course in emergency first aid; such a course includes resuscitation, control of bleeding and treatment of the unconscious person. A small group of up to six can cover this syllabus in two hours' practical training by a nurse or first-aid instructor.

These skills are covered in the excellent 'Save a Life' films first shown on BBC television in the autumn of 1986. These are available on video and can be backed up by a local practical session. Unfounded fears of getting AIDS from the 'kiss of life' has put some people off learning; it is only by using the resuscitation dummy that anyone can be sure of what to do.

Where a full-time occupational health service is provided by the employer it is expected that the first aid arrangements will be made by the doctor or nurse in charge, and so need not adhere strictly to the code of practice, provided they are of at least equivalent standard. Such a service can provide annual refreshers between official certificate courses and be available on the phone to answer queries from first-aiders.

Statutory first aid kit requirements

The employer must provide and maintain first aid boxes. The number of items in a box and hence the size of the box depend upon the number of staff employed. There are five boxes covering numbers up to 150 (see list below). When the number exceeds and is not a multiple of 150, the appropriate kit for the additional number should also be provided. Changes in the regulations are likely to do away with the numbers. To my mind, in that usually only one person at a time is injured, the size two box should suffice in all locations.

Each kit must be strategically situated and readily available. The kit is most conveniently housed in a wall-mounted box, but a satchel should also be provided to take immediate first aid to an injured or collapsed employee.

Any high-risk area warrants its own first aid box, irrespective of other boxes nearby. An example is the kitchen of a catering unit: in the average unit, staff numbers are modest; work starts early – 8 a.m. is usual – before the first-aiders for the rest of the building start work. The accident risk is greater than in an office, from scalds, burns, cuts, and so on. Prompt first aid is essential, so ideally there should be a trained first-aider in each catering unit, but staff may move between units so frequently as to make this impracticable. The next best thing is essential skills training for several staff, given by the company nurse and updated annually. A wall poster 'aide memoire' of first aid treatment is helpful. Good, prompt, effective first aid helps prevent sickness absence, as minor cuts, grazes and burns do not then become infected. Also, freedom from skin infection on the hands avoids one cause of food poisoning.

Ensure the First Aid Box contains sufficient quantities of first aid materials (and nothing else) according to the number of staff, as indicated below:

Ensure that the first aid box is available in a central and prominent position, and is under the control of an 'appointed' person or

Item	Numbers of Staff				
	1–5	*6–10*	*11–50*	*51–100*	*101–150*
'First aid' leaflet	1	1	1	1	1
Individually wrapped sterile adhesive dressings (plasters)	10	20	40	40	40
Sterile eye pads with attachment	1	2	4	6	8
Sterile triangular bandages	1	2	4	6	8
Safety pins	6	6	12	12	12
Medium sterile unmedicated dressings	3	6	8	10	12
Large sterile dressings	1	2	4	6	10
Extra large sterile dressings	1	2	4	6	8

first-aider. Ensure used items are replaced as soon as possible.

A reception area in an entrance foyer should have a first aid kit to be used for staff and visitors; security staff who know the building inside out and are always there, are well placed to be first-aiders. Switchboard telephonists must know who the first-aiders are and where to find them. The name, location and phone number of the first-aider or appointed person must be displayed on notice boards.

Along with the items to be used, and a list of contents, the first aid kit must contain a card or leaflet outlining treatment. This can be designed as a folded card which, besides giving the essential facts about resuscitation, 'breaks', burns, and so on, includes advice on common emergencies and conditions pertinent to that particular industry. In addition, a notebook should be included for recording treatment given and materials used. (This does not replace the accident book held in each department by the Administration Manager.) When items from the kit are used, a named person – the appointed person, first-aider or nurse – is responsible for replacing them as soon as possible from either a central supply or a local chemist. One firm has a computerized central supply scheme for companies with scattered offices, in which each office notifies its needs by postcard; the items are supplied by return, with head office given a periodic overall detailed analysis.

It is sensible that as many staff as possible read and understand the contents of the first aid card (and attend the basic resuscitation course). If the card remains unread until an emergency, precious time is lost working out what to do. When one printed leaflet was replaced by a new edition, we looked for the changes between the two, and discovered an unfortunate printing error in the original, in which the instruction for the treatment of a fractured leg was 'tie the feet and arms together'. No one had pointed this out to us, so it seems likely no staff had actually read the leaflet right through. The notes prepared for the card used for the National Westminster Bank first aid boxes follows later in this chapter.

A first aid box must contain the listed contents only, nothing else at all. All kinds of 'potions and lotions' tend to find their way in and must be discarded. No longer, for example, does the kit include eye ointment. Removal of pain-relieving tablets caused most upset: but a first-aider is trained neither to prescribe nor to dispense, so it is far better for adults to buy their own from a chemist or from a wall-mounted dispensing machine. The no-tablet rule removed large, unlabelled, undated bottles of aspirin of uncertain age which lurked inside many a first-aid box, likely to do more harm than good. A pair of scissors, attached by a chain, is a useful item to put beside the box. A regular, frequent check of the first aid kit may reveal disappearing stock. It would be wrong to keep the box locked and delay treatment whilst the key is fetched, so there is always a chance of pilfering; publicising thefts as anti-social is a deterrent.

Forthcoming changes in the regulations

In the Autumn of 1987, the Health and Safety Commission published a draft revised approved Code of Practice as a consultative document. Revisions are considered necessary because HSE Medical Services Division surveys into knowledge of and compliance with the regulations, have shown a lack of understanding about first aid materials (35 per cent of firms in the survey had in their first aid boxes 'materials other than those recommended by the HSE or provided on the advice of a full-time occupational health service'), and about training for specific hazards (only 8 per cent of firms with a potential chemical hazard had anyone specially trained to deal with it). The opportunity is also being taken to link the amount of first aid provided to work hazards rather than to staff numbers.

The wording of some sections is made easier to understand and further practical guidance is given.

In practice, numbers have been the only factor considered by employers with no thought apparently given to hazards, the area covered, and the distance from emergency services.

Few people have gone on to become occupational first-aiders, so it is felt better to train first-aiders 'in house' to deal with the specific hazards of the particular work place. Specific antidotes can then be kept with the first aid box.

Irrigation of the eyes is now included in training because of the high number of eye injuries, despite the 'Protection of Eyes Regulations'.

Where there is no readily available mains water supply, three containers of at least 300ml each of sterile water or 0.9 per cent sterile saline are required.

The refresher course becomes two days rather than one. This is already sensibly put into practice by the British Red Cross Society and St John Ambulance (who between them provide 80 per cent of first aid training), given the virtual impossibility of cramming comprehensive revision into one day, with half of that day taken up with the examination. In the three years between courses, first-aiders who have not had to practice first aid will have become very rusty so the longer refresher course makes sense.

There is reinforcement of the rule of thumb that if there is a need for a first-aider, two first-aiders should be trained to cover the holidays and sickness absences of either.

There remains a veto on having medication of any kind in the first aid box.

Provision of first aid facilities, equipment and personnel must be 'adequate and appropriate in the circumstances', thus giving an occupational health service very much more discretionary power in deciding what is adequate and appropriate. The present ratios (for provision of one trained first-aider per 150 staff, in an office setting; for provision of a first aid room if more than 400 staff are employed, and for numbers of plasters, bandages, and so on) are to be done away with. The sole deciding factors will be what is considered to be adequate and appropriate – always given that the more people there are in a location, the more likely statistically it is that someone will have an accident or be suddenly taken ill.

In a shift system of working, each shift must be covered.

Employees working in isolated locations or remote areas far from NHS accident and emergency facilities should be trained in emergency first aid.

First aid does not need the use of antiseptics – soap and water suffice – but individually wrapped moist cleaning wipes, free from alcohol, are useful where there is no readily available source of mains water.

The first aid room should have washable walls and floors.

Potential first-aiders must be selected with care. These people should not only be readily available, reliable and able to keep calm in an emergency, they need aptitude and ability and must be sufficiently physically fit to cope both with the first aid course and to react at a moment's notice. First-aiders must be made aware of sources of expert advice.

If oxygen is to be used, additional training is needed over and above the four-day first aid course.

The wording of the notes of the First Aid Leaflet have changed somewhat, but still do not include cardiac massage. I have commented on this, as to my mind this information should be given, as it is covered by the term 'resuscitation', used in the Code of Practice. When breathing has stopped, the heart has often also stopped beating effectively and then, of course, artificial respiration alone is not enough to maintain life.

An individual is encouraged to self-treat a very minor wound but I think that even such minor injuries should be treated by a first-aider, as it is such wounds, poorly treated, which become infected and can cause more problems than a more serious wound which has been properly treated.

FIRST AID NOTES

These notes are taken from those prepared for the illustrated card placed in first aid boxes at the National Westminster Bank:

General

Wash your hands before and after treating minor injuries.

First cover any cut or graze on your own hands or wear disposable gloves.

Dispose of all soiled items safely.

Don't touch the surface of dressings which will be in contact with the wound.

Ensure casualty is either sitting or lying down to be treated.

Report all accidents and ensure entry is made in the Accident Book.

Replace all first aid material you use as soon as possible.

Use protective clothing and equipment where necessary.
Telephone advice available from Occupational Health Centre.

Advice on treatment

Take care not to become another casualty.
Act quickly; be gentle, calm and reassuring.
Remove danger from the casualty. Treat on the spot. Only move a casualty who may be seriously injured if the danger cannot be removed.
Evaluate the injuries by looking, asking and feeling.
Where necessary send for help immediately: first–aider or 'appointed' person in first aid to make a 999 call for ambulance where appropriate, giving a clear message to include exact location, your telephone number, number of casualties and seriousness of their injuries.

Priorities

Before anything else, ensure casualty is conscious and breathing by talking to the casualty. If no response, tug hairs in front of the individual's ear; if still no response, listen for breathing, with your ear against the casualty's nose and mouth. If breathing but unconscious pull, by holding clothing at shoulder and waist, to turn into recovery position, i.e. face down, chin up, head to one side with no pillow under head. Pull up the arm and the leg on the side the head is facing and free the underlying arm. If NOT breathing, start resuscitation by the 'kiss of life' at once. Clear mouth of debris. Bend head right back and push jaw up to relieve obstructed breathing. If breathing does not restart at once, give mouth-to-mouth resuscitation at once – every second counts.
Pinch nose, keep chin up, take deep breath, seal the casualty's mouth with your mouth and breathe out slowly and deeply, watching from the corner of your eye to see the chest rise. Take your mouth away and the chest falls. Two quick breaths, then 15 times each minute. If the chest fails to rise, make sure the head is tilted right back as far as it will go.
After the two breaths, check the heart is beating by feeling the carotid pulse on one side of the neck, with 2 *fingers* pressed

firmly in the groove at the side of the Adam's Apple.

No pulse and ashen or purple face colour, means the heart has stopped beating effectively. (Look at the nail-beds and inside the lips of a coloured person)

Never practise heart massage if the heart is beating.

Kneel beside casualty who is lying on his back on a hard surface.

Feel for top and lower end of breast bone to find the mid-point.

Place the heel of one hand on the lower third of breast-bone and grip that wrist with your other hand, lean forward, arms outstretched, shoulders vertically over the casualty.

Press down firmly and quickly and release quickly at a rate of one a second, and a rhythm of 1 and 2 and 3 . . .

After pressing 15 times, stop and give 2 breaths mouth-to-mouth.

After 1 minute and then every subsequent 3 minutes, check the carotid pulse. Continue until the heart restarts. Two people should work continuously with no stops.

Continue mouth-to-mouth resuscitation until breathing is restored or until a doctor/nurse/ambulance personnel take over.

Once the heart beat and breathing are restored, turn the casualty into the recovery position. Stay with the casualty.

Bleeding

If more than minimal, press on the wound at once with a sterilised dressing pad or clean handkerchief or, if nothing else is available, with your hand. Raise a bleeding limb above the heart.

Broken bone

If a fracture or dislocation is suspected do not move the casualty except from danger whilst summoning skilled help.

Secure the injured part so that it cannot move. Use plenty of padding.

Wherever possible use the other limb or the casualty's body as a splint.

Shock

Any serious injury is accompanied by shock which can be life-threatening.

Reduce shock by a calm manner, reassurance, covering the body to prevent heat loss (but not unduly warming). Give nothing by mouth.

Burns and scalds

Small burns and scalds should be treated by drenching the affected area at once with plenty of clean cool water for 10 minutes before applying a sterilised dressing or clean towel. Where the burn is large or deep, just apply a dry sterile dressing. (NB: Do not burst blisters or remove clothing sticking to burns or scalds.) Remove rings at once from a burned hand.

Chemical burns

Carefully remove contaminated clothing which shows no sign of sticking to the skin and drench all affected parts of the body with plenty of clean, cool water, ensuring that all the chemical is so diluted as to be rendered harmless.

Apply a sterilised dressing to exposed damaged skin and clean towels to damaged areas where the clothing cannot be removed. (NB: Take care when treating the casualty that the chemical does not get on you.)

Something in the eye

Wash out the open eye at once following the instructions on the eye-wash pack. Continue for at least 10 minutes. Then send casualty to hospital.

Electric shock

Ensure that the current is switched off. If this is impossible, free the person, by standing on non-conducting material (rubber mat, wood, linoleum) and using something DRY made of rubber, dry cloth, or wood or a folded newspaper; use the casualty's own clothing if dry. BE CAREFUL not to touch the casualty's skin before the current is switched off. If breathing is failing or has stopped, then give the 'kiss of life' and continue until breathing is restored or medical, nursing or ambulance personnel take over.

Gassing

Move the casualty to fresh air but make sure that whoever does this is wearing suitable respiratory protection. If breathing has stopped, start resuscitation and continue until breathing is restored or until medical, nursing or ambulance personnel take over.

When the casualty goes to hospital send a note of the gas involved with the casualty.

Nose bleed

Sit down.
Head slightly forward.
Breathe through the mouth.
Pinch the entire soft part of the nose between finger and thumb for 10 minutes.
Do not blow nose or remove clots.

Heart attack

Intense, gripping central chest pain typically spreading to left shoulder and arm and sometimes to the jaw and/or right shoulder and arm, coupled with a pale, cold and perspiring skin.
Send for medical help.
Keep lying down, or if breathless, in the most comfortable position.
Loosen tight clothing.
Reassure.

Fainting

If a person feels faint, get head right down – preferably let the casualty lie down flat on the floor.
Loosen tight clothing.
Quick recovery follows.

Epileptic fit

In a fit there is unconsciousness and rigidity, then jerking movements when the tongue may be bitten and the person may be incontinent.

Protect from danger but do not otherwise restrain. Once the jerking has stopped, put in the recovery position. Accompany home to seek medical advice.

Diabetic complication

A low blood sugar causes changes in behaviour.
Person becomes vague, unco-ordinated, pale and sweating with slurred speech.
Give a teaspoon of sugar and repeat after ten minutes.
If unconscious give nothing by mouth, place in recovery position and send for medical help.

Choking

A person who is choking cannot cough or speak (but will nod if asked 'Are you choking?'), and characteristically raises one hand to throat.
Give four slaps between shoulder blades with head low.
If object not dislodged, proceed to give 'abdominal thrust':
Stand behind patient, arms around. Place the thumb of one fist between navel and the lower end of the breast-bone. Cover this fist with your other hand and thrust up and in towards a point mid-way between the shoulder blades. Repeat up to four times and the object should shoot out of the mouth. This action can be carried out on a person who is standing or sitting. If the person is lying on the ground, turn on to back, kneel across thighs, overlap the heels of both hands mid-way between the navel and the lower end of the breast-bone and thrust up and in towards a point mid-way between the shoulder blades. Mouth-to-mouth resuscitation may also be needed. After recovery seek medical aid.

Road crash

Protect the scene of the accident. Stop traffic from both directions with bystanders' signals or warning triangles 100 metres or more ahead and behind the scened. Switch off the ignition and lights of any car involved. Enforce no smoking and no matches. Use the lights of undamaged vehicles to illuminate the scene at night. If there is smouldering from an engine, use appropriate fire extinguishers. Make certain car brakes are on.

Alert the rescue services. Bystander 'phones 999 for ambulance and
  police, stating the exact location of the crash and of the telephone, the number of victims and their apparent injuries.

Give first aid, treating first anyone who is unconscious, or not
  breathing, or whose heart is not beating, or who has severe
  bleeding. Search the area of a crash for anyone who may have
  been thrown out of a vehicle over a low wall or hedge. Treat
  casualties where they are in the car and do not try to pull them
  out; wait for the rescue services.

Sprain or strain

Apply an ice pack or cold pad to the injured joint or muscle.
Avoid tightness which could damage nerves and blood vessels.
Raise the limb.

Bleeding tooth socket

Do not plug the socket itself. Get the patient to bite hard on a thick
  gauze pad or handkerchief for at least 15 minutes sitting with
  elbow on table and hand cupped under chin to give counter-
  pressure.

© NWB 1987

The notes for first aid boxes are only intended as a guide, to supple-
ment practical instruction: when learning what to do, there is no
substitute for practical knowledge from an approved first aid or
resuscitation course. Artificial respiration by the mouth-to-mouth
method and other life-saving techniques can only be thoroughly
learned by practising on a special resuscitation doll, fondly known by
all as 'Resusci Annie'.

FURTHER TIPS ON FIRST AID

Secondary infection is a risk from any breach of the skin and can be
avoided by scrupulous attention to cleanliness. First–aiders therefore
must first wash their own hands before treating the cut, then wash the
cut under gently running cold water, then thoroughly dry it before
applying a dressing. Confidence is rapidly lost if the dressing peels

straight off again. Dirty wounds first need the removal of any oil or grease, then washing with soap and swabs under gently running cold water, using each swab once only, wiping away from the edge of the wound, not sweeping dirt into the wound. Use Sellotape to keep a dressing in place if the casualty is allergic to plaster. Another complication of a deep cut may be tendon or nerve damage, suggested by loss of power or feeling in the area supplied by that nerve. The first–aider must also always be on the lookout for something in the wound, such as a piece of glass, and if this is suspected, refer to hospital. Any risk of AIDS to first–aiders is overcome by first covering any cut or graze on their own hands with a waterproof dressing and washing their hands after giving treatment as well as before – or by wearing disposable gloves (see the later section on AIDS in chapter 15).

A splash of chemical into the eye or on to the skin is treated with prolonged copious water application. Anywhere chemicals could splash into the eye, in a battery acid store, for example, running tap water or a saline pack must be readily available close by. Saline is best contained in a rigid, disposable plastic bottle with simple, clear instructions taped to the outside. The cap is twisted off, which may involve tightening it first to break the seal; unless realised, this may exasperate people and delay when every second counts. The casualty lies down and whilst one person pours, a second holds the eyelids apart, as the pain causes muscular spasm, to allow the fluid to flood the surface of the eye. Irrigation must go on until the victim is in expert hands and for at least ten timed minutes. After use the bottle is disposed of and replaced. The expiry date on the bottle should be adhered to. Best stored in a cool, dry, dark place, the bottle should be inspected periodically to check it remains full and clear.

When there is risk of eye injury from a foreign body, chemical splash, and so on, protective goggles must be worn. If the employee thinks something went into his or her eye, even when nothing can be seen by the first-aider, it is essential to refer that person to hospital. For instance, a sharp metal fragment which penetrates and lies within the eye may leave no obvious mark on the surface of the eye.

COMPANY FIRST-AIDERS

Some employers pay staff who carry out first aid duties. This may be either a single payment when the first aid exam is passed with a ceremony to hand over the certificate being worthwhile, or an annual payment. This could be based on the degree of responsibility under-taken, with category A responsible for all first aid, category B regularly deputising for example at lunch breaks, category C a back up to the A and B first-aiders to cover annual leave, sick leave, and so on – but there could then be the situation of a category C first-aider being at the scene of an accident and not allowed to provide treat-ment because the category A or B first-aider has to be sent for. Better I think, if payment is to be made at all, is a flat rate for all active first-aiders, with possibly a retainer for other trained first-aiders not required for the time being because of numbers or location.

Offices are reasonably safe places to work, and first-aiders may have little or no opportunity to practise the skills they have learned in the three years between refresher courses. To maintain interest and standards some companies organise competitions between teams of three or four first-aiders. Their efforts to deal with simulated accident situations, using Casualties Union volunteers, authentically made up, are judged by two doctors. Friendly inter-department or inter-company rivalry is encouraged and at the end prizes are presented to the winning team after constructive comments by the doctors.

Procedural difficulties sometimes arise when a company nurse is first appointed; first-aiders who have been 'in charge' until then may be reluctant to accept the nurse's authority. Their roles however are quite different and first-aider cover needs to continue after the appointment of a nurse; the presence of a nurse does not mean one can dispense with first-aiders. First-aiders should continue to provide immediate care at all times.

More first-aiders are needed to cover shift work or flexitime. The employer should also consider providing fully trained first aid cover for nights and weekends, even when numbers are below the usual figure, if a substantial number of staff are at work, or when the work is potentially hazardous and there is no nearby hospital accident and emergency department.

Staff can expect too much of first-aiders and regard them as substitute doctors. Some employees may seek advice about chronic conditions and anyone with a recurrent symptom like a headache should be directed to their own GP. A first-aider is taught not to exceed the brief of only giving first aid, and will refer if in any doubt. When referring an employee to the care of a GP or hospital, a letter containing a full statement of the event and treatment given should also be sent.

Detailed records of all treatment given must be kept by the

first-aider or the person in charge of a first aid box. From this information, some analysis of ill-health and of injuries can be made. The first-aider takes charge at the scene of an accident, and must not allow himself or herself to be browbeaten by the presence of senior staff, but instead should exert his or her authority.

INSURANCE COVER

Although in practice it is rarely needed if at all, company insurance cover is necessary to cover the possibility of a maltreatment claim against the employer's first-aider, alleging that incorrect treatment has caused prolonged suffering and/or caused disability. Provided the first-aider has acted skilfully in the best interests of the accident victim, like a 'Good Samaritan', such a claim would not succeed.

Members of the British Red Cross Society and St John Ambulance are covered by the insurance policies of these organisations when on duty for them, but this cover does not extend to the first-aider at the work place. Nurses who are members of the Royal College of Nursing are insured by the RCN.

THE FIRST AID ROOM

Ideally a company first-aider should have use of a first aid room and at present such a room must be provided when over 400 staff are employed. The size depends on numbers to be treated and therefore on the size of the establishment. The minimum floor space is 100 sq. ft. Essential are: a telephone, a sink with running hot and cold water (the cold tap to be drinking water), soap and a nail brush, a smooth-topped work surface, adequate light, a refuse container, a chair, a couch with space around it and pillows and blankets which are regularly changed and frequently cleaned, a carrying chair and of course first aid equipment. A toilet should be nearby. The room must be cleaned every day, even if the room has not been used (and it is all too easy to overlook this). The floor should be vinyl; screens should be provided to give a degree of privacy and to enable the treatment of more than one patient at a time. Blinds allow the room to be darkened for an employee to rest.

If a properly-equipped first aid room is provided, fewer first aid kits are needed. All treatments must be carried out in the first aid room which must be used for nothing else (other than as a doctor's consulting room).

A first-aider must always be readily available during working hours and must ensure that supplies of first aid equipment, dressings, and so on are properly maintained at all times.

Suggested check list for first aid room

If possible the above room should be:

(1) Situated on Ground Floor and near Exit.
(2) (If above ground level), situated near a lift.
(3) Have emergency lighting available.
(4) Have toilet facilities in near vicinity.
(5) (If waiting area available), chair in this area.
(6) Blinds at windows (vertical blinds can be kept clean more easily).

The following items are requirements:

Sink with washable working surface.
Soap & nail brush.
Notice on door listing first-aiders, their locations and phone numbers.
Vinyl sheet flooring, with coved edges.
Drinking water and disposable cups.
Telephone.
Notice board.
Store cupboard for 2 blankets and pillows (blankets, cotton cellular; pillows to have plastic cover and disposable cases).
Paper towels with wall holder.
Wall thermometer.
Chair for patient.
Chair for first-aider.
Sealer unit for disposing of soiled dressings.
Bags for above item.
Examination couch.
Bed chair.
Wall-fitted First Aid Box.
'Normal' saline sterile eye wash.
Bucket with lid.
Clinical thermometer.
Plastic aprons and disposable gloves.
Carrying chair.

5 Specialised Health Facilities

Special leave is usually given for a visit to a health care professional, be this a GP, dentist, hospital specialist, optician, physiotherapist or chiropodist, as it may only be possible for patients to be seen during office hours, sometimes involving the loss of half a day's work. If the time which would otherwise be spent at work, including travelling to and from the appointment as well as the visit itself, is costed, the cost benefit of some arrangement for local or 'on-site' provision of some of these facilities will be worth considering, especially for a City-based company where the staff commute from the suburbs and beyond, which is where the health care professionals they consult are likely to be based.

DENTAL SERVICES

In a few countries, an occupational dental service is mandatory, or agreements with trade unions include this provision. A few UK companies provide in-house dental treatment because the good dental health of their employees is relevant to their work; with food producers and handlers, for example, dental sepsis may not cause food poisoning but it does not inspire consumer confidence, and is a pointer to lax personal hygiene which may contribute to food poisoning. The manufacturing process may also affect the teeth or mouths of employees. Besides regular visits for routine inspection, most adults at some time need treatment for tooth and gum disorders.

In the United Kingdom the General Dental Service (GDS) of the NHS is responsible for providing dental care. As mentioned, it can be very time-consuming for an employee who works many miles from home to be treated near home. Appointments outside office hours may not be available and the cost of practice accommodation in city centres has meant fewer dental practices offering comprehensive NHS dental treatment there. In one city centre as few as 10 per cent of dentists offered the full range of NHS treatment, because of the slender profit margin for treatments.

The dentist is usually paid per item of treatment and has low profit

51

margins; the answer is a high throughput of work to remain viable. In the NHS this work rate is the highest in the world.

Dental treatment is not synonymous with dental care, which includes steps to prevent tooth decay and gum disease. Preventive care is partly educational, takes time and is not cost-effective.

With an 'in-house' dental service, the employee is away from work much less and gets a comprehensive service with emphasis on prevention. Provided the employee wants it, a healthy mouth can be achieved and maintained with, in time, minimal involvement by the dentist.

Avoiding dental emergencies helps maintain work flow, especially where the employee is a key member of a team. Such emergencies cannot be eliminated altogether but they can be decreased considerably by regular dental care.

Provision of dental services for UK occupational groups started in the early 1900s. Since the start of the NHS there has been unique co-operation between the public and private sector; the only proviso is that an occupational dental service is for employees only.

There are several ways this service can be run, but the best I think is where the dentist is a salaried employee of the company, registered as an independent dental practitioner with the Family Practitioner Committee of the NHS and subject to the terms of service of the GDS. Staff attending as patients pay their contributions towards the cost of the treatment as specified by the NHS regulations. These fees are passed on, by agreement, to the employing company and offset the costs of the service. Thus with economic restrictions lifted, there is more time for high-standard dental care. This creates a framework for a first-class service.

The dental surgery can conveniently be accommodated adjoining the company doctor's rooms in the occupational health centre, with a shared reception and waiting area. The capital cost of the specialised equipment is substantial and under-floor plumbing is necessary. A trained dental nurse as a 'chair-side' assistant is required.

Ancillary help from a dental hygienist skilled in teaching preventive techniques is invaluable. A hygienist may not be employed without a dentist. The engagement of a full-time dentist, not normally worthwhile unless there are at least 1500 employees, depends on such factors as: the rate of turnover of staff; the number without their own teeth; the availability and quality of local dental services; any special responsibilities to staff under training or expatriates; specific occupational requirements; the employees' wishes, and so on. The Association of Industrial Dental Surgeons (Royal College of Surgeons of England, Lincoln's Inn Fields, London), will advise.

A retail company with many small units distributed throughout the country cannot provide the above service, but several do provide the facility of an 'in-house' dental inspection twice a year by a local dentist. Staff are told whether they need treatment, which will be by the dentist of their choice, and preventive advice is given by a dental hygienist.

Alternative arrangements for engaging a dentist differ in their methods of remuneration and also vary with the degree of involvement of the dentist with the company. An extreme case is the dentist who provides an NHS practice on company premises, but retains all NHS fees.

An 'emergency only' service has appeal, because this is the prime dental reason for unplanned absence from work, and one dentist can cope with several thousand employees. Such a service is unlikely, however, to lower the prevalence of dental disease or the incidence of dental emergencies, and the limited nature of such work is unattractive to most dentists.

OPTICIAN SERVICES

Another paramedical service which can be arranged on site is a regular visit by an optician. There are several categories of optician.

A dispensing optician makes up the spectacle prescriptions of other opticians.

An opthalmic optician or optometrist who is identified by the letters FBOA (Fellow of the British Optical Association) is trained in sight-testing, the prescribing of spectacles and recognising abnormalities which require referral to a doctor (a number of medical practitioners with the relevant diploma, DOMS, are also trained in sight-testing and prescribing). A consultant in this field is an opthalmologist who will be a Fellow of the Royal College of Surgeons (FRCS).

Many opticians now have the equipment and training for testing for silent glaucoma by measuring eyeball pressure with a controlled, painless puff of air against the eye.

PHYSIOTHERAPY SERVICES

Some NHS Physiotherapy Departments have waiting lists and this is one reason why a number of companies provide 'in-house' physiotherapy. Another reason is the significant amount of time off work needed for what is usually a course of treatment given two or three times a week, plus the time spent travelling to and from a hospital physiotherapy department which can involve a half-day lost from work for each appointment.

Prompt physiotherapy facilitates early and total recovery from an injury such as a sprained ankle, so the provision of a physiotherapy unit may allow an earlier return to work.

The physiotherapist, a Member of the Chartered Society of Physiotherapists (MCSP), who is likely to be a member of the Association of Chartered Physiotherapists in Industry, may be employed full-time by a large company or part-time on a sessional basis. There should be at least three sessions in the week, as most conditions suitable for physiotherapy respond best to treatment which is prompt, regular and frequent.

A course of treatment is given at the request of a GP or specialist, with the physiotherapist usually deciding what treatment is appropriate. There is no self-referral but a company nurse in an Occupational Health Service may refer via the company doctor for assessment and first treatment, with GP sanction sought for continuing the course. The company doctor will also see the patient who fails to get better in a reasonable time before referral back to the general practitioner, who may then refer on to a specialist. The emphasis of the unit is on early treatment of acute conditions. Treatment in the unit may be supplemented with exercises to perform at home plus advice on how to hasten recovery and prevent recurrence by correct posture, correct methods of lifting and so on. Appointments for the unit must not be filled by those with chronic disorders which no amount of physiotherapy will cure; after a course of treatment, a person with such a condition should be told this.

The image of physiotherapy as passive, with soothing heat and massage, is incorrect. The emphasis is on active treatment and involvement of the patient. Often advice as well as treatment is needed, encouraging the person to self-help with instruction in correct methods of lifting, moving, sitting, and so on. A leaflet can be used for this purpose but personal instruction is best.

The physiotherapy unit need not take up much space. The types of equipment to be installed should be discussed with the appointed physiotherapist before work starts. (Delivery delays must be taken into account.)

Suggested equipment could include machines for:

interferential
short-wave therapy
traction
ultrasonics
faradism.

Also useful:

refrigerator for ice packs
traction couch
'wobble board' for ankle exercises
barbell weights
medicine balls
splints
bandages
skeleton to demonstrate anatomy.

The unit needs:

(1) Nonslip floor.
(2) Wash-hand basin.
(3) Electrical sockets at waist level.
(4) Paper sheets, paper towels, pillows and blankets.
(5) Hooks for pulleys.
(6) Hydraulic couches for manipulative techniques as well as other forms of treatment.
(7) Cubicles, screens and curtains to give patients privacy.

Conditions most often treated by physiotherapists are:
Low back pain 44 per cent Neck pain 27 per cent Joint disorders 16 per cent Sports injuries 13 per cent. Very careful assessment is needed before back pain is treated, because there are types of back pain for which physiotherapy is inappropriate.

CHIROPODY SERVICES

Providing subsidised chiropody at work is popular with staff, but is not that essential. Foot disorders may disable, demoralise and contribute to loss of efficiency, (but rarely cause sickness absence). Many people could do more to look after their own feet.

Many accidents at work are caused by slips and falls, in some cases contributed to by the shoe style or by uneven or excessive wear of the shoes. These are subjects on which a chiropodist can advise but are equally the province of the safety officer.

NHS chiropody is available in pregnancy, for the elderly, and for those whose feet are particularly at risk because of certain medical conditions in which they cannot feel their feet, such as complications of diabetes. Some foot disorders, such as an ingrowing toenail or a bunion, are also treated by a surgeon; verrucas are also treated by a GP or dermatologist.

One full-time private chiropodist covers 10 000 of the population. This figure cannot be directly compared with industry, as it includes all age groups, and whilst few children and only a tiny proportion of adults need chiropody, many old people do, as it becomes more difficult for them to look after their own feet.

Much treatment relates to removal of corns and patches of hard skin on the feet, and the feet are more comfortable when these have been removed. Often certain styles of shoe have caused the problem and unless these are not worn, the problem will recur time and again and need regular treatment, every two or three months. Thus a chiropody service can get filled up by relatively few people.

The service may be restricted to staff on their feet for much of the day, for instance, messengers, cleaners and catering staff. It is important their feet are comfortable but most should be able to care for their own feet and after all, feet are designed to be used!

A chiropodist will be both MChS – a Member of the Society of Chiropodists, and SRCh – State Registered Chiropodist. There has been a shortage of State Registered Chiropodists. A company may use the services of a local chiropodist on a sessional basis. In one session lasting three hours, treatment can be given to both feet of seven people.

Provided by the company are the room, light, heat, hot water and so on together with the necessary equipment, lotions and dressings. Equipment required includes:

(1) Reclining chiropody chair on hydraulic base (manual or electric) with adjustable footrest. The chair is heavy and needs a fixed location. Any other chair is unacceptable.
(2) Operator's adjustable, washable chair on castors.

(3) Dressing trolley with spool rail.
(4) Drill incorporating dust extractor (for treating some nail defects).
(5) Anglepoise light.
(6) Autoclave to sterilise instruments.

A chiropodist is qualified both to diagnose and prescribe without reference to a doctor. An appointment may be made directly with the chiropodist. A chiropodist can advise on foot hygiene and the prevention of smelly feet which can be offensive in the summer. Advice is also given about footwear for specific situations, such as in the catering department, where there is a risk of scalds from spillage of very hot liquid or fat on the foot of a worker wearing sandals. As mentioned earlier, the chiropodist will remind catering staff of the importance of wearing shoes with enclosed toe pieces and nonslip soles.

The chiropodist will advise management on the type of reinforced safety shoes to provide when there is a risk of crush injury or of exceptions where a variation should be allowed - for instance where an individual foot problem needs leather rather than steel-reinforced toecaps.

INFLUENZA VACCINATION

Influenza puts in a regular appearance as a reason for self-certificated sick leave for absences of a few days; but anyone who has suffered real influenza knows that two weeks are needed for recovery. Influenza is thus an over-self-diagnosed condition: it is very different from a common cold. It presents with sudden onset of fever, shivering, headache, aching limbs and 'feeling rotten', followed by a cough and sore throat. When these clear there can be lingering lassitude, fatigue, general debility and depression. In the acute stage, bed is the only place to be, with plenty to drink and regular pain-relieving tablets.

Except in an epidemic, for every true case of influenza, there are about ten cases of other, lesser virus infections. Colds, other respiratory tract infections and bronchitis are also often called influenza. Influenza spreads west each winter from the Far East. It is usually caused by a type 'A' virus, less often by type 'B'. From time to time these viruses change slightly and a vaccine protecting against previous strains is then largely ineffective. The vaccine is made from up to three varieties of the killed virus which is thought by the World Health Organisation to be most likely to strike that year. Within two weeks of being given a dose of vaccine, there is protection for at least six months against those three viruses: Natural infection protects against reinfection for three years. It is common for people to consider the vaccine ineffective if soon after the vaccination they develop a cold. They think the vaccine either caused the cold or has failed to protect them and do not realise that the vaccine only protects against influenza, and against the specific strains of virus in the vaccine that year. Nowadays the vaccine is not given to healthy adults, but only to those with an existing chronic lung, heart or kidney disorder or those in certain vulnerable groups, such as the elderly in residential homes. It used to be recommended for those in key jobs who must be at work at all costs, but this is no longer the case, and fewer healthy adults want it anyway, although a few remain convinced of its benefit. It was recommended by one vaccine manufacturer to keep the wheels of industry turning. One organisation conducted a five-year study into the cost-effectiveness of flu vaccine offered annually to 60 000 staff and found less sickness absence in the winter months. During these months 30 fewer working days were lost through sickness absence in one department of 100 people than in a similar department where the vaccine was not offered. Doctor/nurse teams are no longer sent to administer the vaccine for a set fee; the jet gun they used has been abandoned because of the risk of transmitting AIDS. Nowadays, a pre-loaded single-dose syringe is used. Last-minute loss of nerve and genuine sickness explain the disparity between the number who say they will have the vaccine and the number who actually turn up for it.

The vaccine is grown on chicken eggs and so is not offered to the very few people who are truly allergic to chicken eggs, meat or chicken feathers. Rather more people simply do not like, and so do not eat, eggs or chicken meat; this is not the same as an allergy. The vaccine is given by injection and this is another reason for nonacceptance: no one likes an injection.

Side effects usually consist of temporary redness and soreness at the injection site. Occasionally there is a generalised reaction of fever, aching muscles and malaise for a day.

The ideal vaccine should be live, inhaled and longlasting. The lining of the nose is where the influenza virus acts and so where immunity should be stimulated. Such a vaccine has yet to be perfected.

BLOOD DONATION

In recent years the need for blood donors has increased, and at times the shortage of blood has been desperate, as can be seen from posters exhorting us all to give blood. Hospitals, through the work of the National Blood Transfusion Service (NBTS), depend on volunteer donors aged 18–65 for supplies of blood and blood products; of those eligible to donate blood, only thirty per thousand do so. Companies can contribute by twice a year allowing staff to attend to give blood during working hours, either at a nearby donation centre or at a mobile unit visiting their building. In large cities, such as Manchester and London, there are NBTS centres where anyone, local worker or passerby, can attend, preferably by appointment and, if there are no medical contra-indications, donate a pint of blood. Temporary contra-indications to giving blood include:

(1) Recent vaccination and/or overseas travel.
(2) Current illness.
(3) Anaemia (this is always routinely tested for before the blood is taken and if found the person is advised to see their own GP).

Other contra-indications are:

(1) Epilepsy.
(2) Diabetes.
(3) History of jaundice (for twelve months after recovery).
(4) History of malaria (though plasma can be used).
(5) AIDS and carrying HIV, the AIDS virus.

For large Companies, the NBTS collect the blood on site by sending a team of a doctor, nurses, technician and receptionist.

The NBTS provides advance promotional posters, overstamped with details of the sessions, and supplies all equipment, beds, pillows, blankets, and so on. The 'host' company provides chairs and tables in a readily accessible large room with nearby toilets. The room needs to be pleasantly warm but not stuffy, and well lit. Efficient ventilation and good lighting are therefore essential, plus several power points.

The company arranges the appointment timetable: six employees can be dealt with every quarter of an hour. On this basis up to one hundred and thirty are 'processed' a day.

The company 'loses' the time the employee takes to get to and from the donation centre, give a pint of blood and recuperate with a cup of tea. With an efficient appointments system this totals about forty-five minutes.

Blood donors are thanked by personal letter and are given a certificate. With regular donations, they qualify in time for the first of a number of distinctive lapel badges or brooches. Surburban evening donor sessions cater for staff unable to attend in the daytime.

In recent times, people may have been put off giving blood because of AIDS. This fear was voiced in a question at a 1986 public meeting on AIDS, which was addressed by an expert from St Mary's Hospital, Paddington. She reassured her listeners that there is no risk whatsoever to a blood donor from the Human Immunodeficiency Virus (HIV), as all the equipment used is absolutely sterile (unlike the equipment of a drug addict who may use a needle already used by another addict who is harbouring HIV). Each donor's blood is now screened for HIV and if this is found and confirmed by more elaborate tests, to avoid any chance of a false positive, the donor is told and counselled and the blood is not used for transfusion. There is therefore now virtually no risk to anyone in this country of acquiring HIV from a blood transfusion.

THE OCCUPATIONAL HYGIENIST

In large companies and in those with particular potential hazards, another member of the occupational health team is the occupational hygienist, who is likely to have a science degree such as an MSc and/or a certificate or diploma from the British Examining and Regulation Board of Occupational Hygiene (BERBOH), such as the Dip Occ Hyg, and is likely to be a member of the British Occupational Hygiene Society and to have industrial experience. He or she will contribute to the work involved in prevention of risks to health from the working environment, including investigation and monitoring. Smaller companies which need such investigations can call on the services of a skilled hygienist from a number of private occupational health organisations.

THE OCCUPATIONAL PSYCHOLOGIST

A few companies employ an occupational psychologist who has a degree and is a member of the British Psychological Society. The psychologist's involvement includes study of the effects of work on health; job satisfaction, and peoples' behaviour at work. The occupational psychologist usually also has clinical skills and can give practical help to staff suffering from the effects of stress and other emotional problems. An alternative approach is to use the services of an independent psychologist and this can form part of an Employee Assistance Programme (EAP) in which the patient will have all the more confidence that anything said is confidential.

6 Health Checks and Screening

RECRUITMENT OF NEW STAFF

Recruiting formalities usually include an enquiry about health. Before World War II, to gain employment there was often the requirement to be a 'first-class' life insurance risk and to take out a policy to prove this.

In practice, to be a first-class risk today does not guarantee good health tomorrow, but though such a condition of employment is less apparent nowadays, even in companies where, as regards the work, physical fitness is not critical, stringent guidelines still apply.

Dr John Todd, writing in the *British Medical Journal* on the subject some years ago, referred to a newspaper item headed 'Healthy Man Who Failed Medical, Loses Job'. The man concerned had got a job as a porter and bench hand, but the company doctor found he had hypertension and commented, 'I am sure he is fit for work, but we are not so much concerned about his present health as his future health'. If this kind of medical rejection can be justified by one doctor working for one corporation, it can equally be justified by other doctors working for other corporations. In consequence, unfortunates with, say, hypertension, diabetes, heart murmurs and albuminuria may be unable to get any employment.

Dr. Todd continued:

When a prospective employee is applying for a job, such as an air line pilot or engine driver, in which his sudden incapacity can hazard the lives of others, there is of course no ethical objection to the doctors recommending his rejection if he is found to have a condition making him liable to such incapacity. But many people applying for jobs where this risk to others does not arise, are nevertheless compelled to have a medical examination. And although they have all the right qualifications and are accepted 'subject to medical examination', they may later be informed that because they have 'failed' the medical they cannot after all have the job.

As Dr. Todd implies, there is no need for the employer to turn down

62

an applicant who is in other than perfect health unless the health problem would make the job hazardous for the applicant, his or her colleagues, or the public, or – as the International Labour Organisation states – the proposed job would be harmful to his or her health. While there are some occupations which demand supreme physical and mental fitness and it is reasonable that the employer seeks evidence of this, for clerical work a routine pre-employment medical examination (PEME) is unnecessary.

It is reasonable and sensible for the employer, mindful of the need for efficiency and continuity of the company, to ask about health and, guided by the company doctor, to ensure the applicant is medically fit to perform the tasks expected of him or her, and that the job will not exacerbate an existing condition.

In practice, the range of detail asked of the applicant varies from a single question at interview, 'Are you in good health?' to a detailed technical questionnaire or a comprehensive medical examination by the applicant's GP or the company doctor.

Fewer people now spend their working lifetime with the same company. Besides, it is only practicable for a doctor to assess the likelihood of repeated or prolonged sickness for the next few years and not to try to predict beyond this.

PEME is not foolproof, as present medical fitness does not guarantee future good health. School and work references may give a better guide, certainly as regards short-term sickness absences, as the pattern of these, once established, is usually permanent.

Where the answers to a questionnaire like that below suggest a significant medical problem to the company doctor who receives this information direct from the applicant, it is often better to seek details from the GP than arrange a PEME. If a form like this one is used, the applicant, by completing and signing the form, consents to an enquiry under the Access to Medical Reports Act, details of which as regards an applicant's rights are given on the front cover. The recommended fee for such a factual report from existing records is about half that for a full PEME. If the answers to the questionnaire are all 'no', there is no need to involve the company doctor.

Our experience shows no advantage from a PEME for 'late entrants'; for instance, persons aged over forty, even though statistically they are more likely to have had significant ill-health: PEME is always needed for expatriate staff and their families going overseas (the children being checked by their GP), and for drivers, manual workers and food handlers. This can perfectly well be carried out by

Company Health Facilities

Pre-employment Health Information – In Confidence

Proposed Start Date:

Originating Office:

Reference:

Date:

Applying for: (Please tick as appropriate)

Shift Work	Maintenance
Clerical	Branch Guard
Secretarial	Bullion
Catering	Computer Staff
Messenger/Driver	Others/Specify

---

1   **Surname**

Mr/Mrs/Miss/Ms

First Names:

Date of Birth

Day   Month   Year

---

2   **Private Address:**

Post Code:

Home Telephone Number:

3A **Name and address
of present general
practitioner:**

Post Code:
Telephone No:

3B **Name and address of specialist** (if
applicable):

Post Code:
Telephone No:

Hospital Ref No:

---

4   **Height:**

ft   ins
        cms

Weight:

st   lbs
    kg

---

5   **Please mark Yes/No as appropriate and give full details overleaf.**

Have you ever had an illness, injury or
operation that has kept you away from
school or work, and/or needed hospital
treatment for more than 3 weeks?

Yes/No

If Yes, please give details eg
(when/diagnosis/duration/medication)

Do you have any family history of
disease? (please specify)   Yes/No

Have you had, or do you have, any of
the following? If yes please give details

Heart Disease	Yes/No	Bowel disorder	Yes/No
Angina	Yes/No	Diabetes	Yes/No
Poor circulation	Yes/No	Kidney/Genito-urinary	
Recurrent Asthma	Yes/No	disorder	Yes/No
Recurrent		Recurrent/Chronic skin	
Bronchitis	Yes/No	complaint	Yes/No
Serious head		Severe Hay Fever	Yes/No
injury	Yes/No	Back Pain	Yes/No
Epilepsy	Yes/No	Arthritis	Yes/No
Mental Illness	Yes/No	Multiple Sclerosis	Yes/No
Depression	Yes/No	Deafness	Yes/No
Anxiety State	Yes/No	Speech/Writing defect	Yes/No
Ulcer	Yes/No	Severe Migraine	Yes/No

Have you ever been refused employment for medical reasons? Yes/No

---

6   **Eyesight**

Do you have problems with your eyesight,
even with spectacles or lenses (if normally
worn)?   Yes/No

Are you colour blind?   Yes/No

7A   **If you have answered 'Yes' to any questions in sections 5 and 6 please
return form direct to:**

Occupational Health Department

7B   **If you have answered 'No' to all questions in sections 5 and 6 please
return from direct to:**

Originating Office

---

8   I agree to my doctor submitting a medical report, to the Company's
Medical Adviser, so that advice can be given to the company.
I wish/do not wish to have access to the report before submission.
(Delete as appropriate.)

Signature _____   Date _____

the company nurse, who can 'pass' but not 'fail', with any doubtful fitness cases being referred to the company doctor. When an outside doctor does the PEME, the completed form is sent by that doctor direct to the company doctor. Management are then advised on a form whether or not that person is medically fit for the proposed job. The basis for this advice is not revealed, as to do so would breach medical confidentiality. If there is a condition the employer needs to know about, the recruitment department is asked to alert the company doctor when the person concerned joins the company. The doctor then advises that person to inform their manager, in order to ensure appropriate first aid in an emergency, such as an epileptic fit, whilst at work.

TB in now uncommon, so it is unnecessary to have a routine chest X-ray or any other special investigation.

If employment is offered 'subject to satisfactory medical and references', and the company doctor considers an applicant medically unsuitable for the particular job, and if this advice is acted on, the applicant may deduce that he or she 'failed' the medical and may dispute this. Where the medical background has, with due consent, been divulged by the GP or where this doctor carried out the medical examination, the doctor – patient relationship can be strained, even though it was the company who made the decision about employment. The Access to Medical Reports Act adds a new dimension.

It is therefore ideal that employment is not offered, and that the employee does not resign from an existing job, until all formalities are completed. If health or references make that person unsuitable for the particular job, rejection is then fairly because of competition with others.

The examining doctor, if other than the applicant's GP, will inform the GP about any significant findings and will tell the applicant to consult his or her GP.

Psychometric testing for personality traits is not widely used in this country: managers still rely on their own judgement.

In the past, a common cause for rejection was obesity. In the long-term there is increased risk of a number of illnesses, and a life insurance premium will be 'loaded' accordingly, but a plump school-leaver should remain in as good health as someone of normal weight for at least the next fifteen years, apart possibly from an increased accident risk. Particularly as fewer people now spend all their working lives with one company, weight limits can, for most jobs and for the young especially, be very tolerant: within 35 per cent of the ideal weight for someone's height is a reasonable limit for someone over forty. Why the fat person became fat is a question which should be answered. A 'glandular' defect is unlikely: 'it runs in the family' usually means faulty eating habits. Overeating may have been a reaction to anxiety and worry or mental illness, when food was a solace and comfort. Obesity in a person over forty is more significant, as risks of illnesses causing sickness absence are that much higher than for a lean person. Height and weight need to be verified if there is a medical examination, as the obese tend to overestimate height and underestimate weight.

EMPLOYMENT OF THE MENTALLY OR PHYSICALLY DISABLED

When considering the employment of a known disabled person, each case should be looked at very carefully. The individual, the health problem and the proposed job should be considered by a team made up of management, the company doctor, the GP and the specialists involved, plus the personnel manager for disabled staff where there is such an appointment.

Whether to employ a person with a physical disability, such as the absence of a limb, is a straightforward decision which depends on the degree of disability and the possibility of adapting the workstation and the facilities available at work. Deciding whether to employ a person with a psychological disability is more difficult, as is whether to employ someone suffering from a recurrent or chronic illness, the long-term prognosis of which is guarded. One reason often given for rejection used to be that such a person is unacceptable to the company pension fund, membership of which used to be a condition of employment. But in a large company any deficit to the pension fund from the early retirement on health grounds of people taken on with a disability would anyway have been more than balanced by the majority retiring normally at between ages sixty and sixty-five.

Failing to disclose a significant medical history when directly asked weakens any future case for unfair dismissal to an Industrial Tribunal, should termination result when the truth comes out, as can happen if there is recurrence of a condition which otherwise is outwardly undetectable, such as epilepsy, diabetes and some types of mental illness.

Medical examination for a job which involves using the back, for instance, being a porter, may detect a back problem, and for an applicant with a history of severe or recurrent back pain the company doctor is likely to recommend medical examination and will seek medical details from the GP.

In any claim for industrial injury benefit because of back injury caused by work, a past history of backache is significant and could affect the employer's liability.

Mental illness

A single, short, acute episode of a mental illness in the past should be no bar to employment; there may, for example, have been over-whelming stresses at the time. On the other hand care is needed when there is a history of recurrent episodes of mental illness. This is likely to be due to recurrent anxiety and/or depression in a vulnerable personality, or to schizophrenia recurring despite long-term prophy-lactic treatment, or when, as sometimes happens, treatment has been stopped against medical advice because the person feels well and so does not see the need, or has side-effects, or just does not like medication. Schizophrenia, as well as some varieties of depression, may recur for no apparent reason, that is, without any precipitating stress.

Between episodes, the individual with a recurrent mental condition appears to be and is well, and a PEME will be normal. Hence the value of a self-administered health questionnaire for applicants and for the need for the prospective employer to incorporate in the form a request for the applicant to consent in writing to an enquiry by the company doctor to the GP. There can then be liaison between the company doctor and the applicant's GP, subject only to the Access to Medical Reports Act. The company doctor will then give an opinion on suitability for appointment.

Less serious but by no means less disabling neurotic illnesses recur in susceptible personalities. Here the precipitating stress can usually be determined. (See Chapter 17, 'Aspects of Mental Health'.)

Mental illness and mental handicap are common. One person in ten in the United Kingdom will need hospital treatment at least once in their lifetime for a pyschiatric disorder. Though no longer re-garded by the public with the abhorrence of fifty or more years ago, when those with serius psychiatric disorders were incarcerated in large, remote institutions, mental illness can still carry the stigma of 'weakness' and disclosure of such a history to a prospective employer does risk discrimination. Risk of recurrence is one reason the employer may be wary. Other reasons may be past experience with a mentally ill employee, where the illness and absence lasted many months, or the failure to appreciate the very wide spectrum of mental illness. Alert to this wariness, and perhaps after a number of rejections, the applicant may withhold such a history. (Serious mental illness will leave gaps in a curriculum vitae.)

The National Association for Mental Health, MIND, considers that people who have suffered mental illness need legislation to protect them when they recover and try to find work. Their report, *Nobody Wants You*, details many case histories of people discriminated against because of a history of mental illness.

They consider that all former psychiatric patients should have the legal right to be silent about a history of psychiatric illness. In the USA there is legislation which absolves someone with a history of psychiatric illness from having to disclose this to an employer if a consultant psychiatrist, after individual assessment of mental capacity for employment, regards that person as mentally fit.

Physical disablement

There are sadly many disabled people in the United Kingdom. 'Handicapped and Impaired in Britain', a study from the Office of Population Census Studies, reports that three million physically-impaired people aged over sixteen, two-thirds of them female, live outside institutions, and of these half are handicapped. Arthritis, stroke and bronchitis figure high in the list of causes of handicap; of course, many of those handicapped by these disorders will be over retirement age.

In connection with disablement, mention should be made of the Job Release Scheme, a Government measure to reduce unemployment. The first scheme covered men of sixty-four and women of fifty-nine in full-time employment. The scheme was then extended to include men aged sixty-two to sixty-four and disabled men aged sixty to sixty-four. Disabled men are defined as those who, on account of injury, disease or congenital deformity, are substantially handicapped in obtaining or keeping employment suited to their age, qualification and experience. These people have the opportunity to apply to retire before the statutory pensionable age, provided their employers agree. The employer must recruit a replacement as soon as possible from the unemployed register, not necessarily to the same job. A disabled man, if not registered as disabled, submits with the application form a medical report form completed by his GP. If a disabled worker is retired under the scheme (which has now stopped), the replacement must (whenever possible), be a disabled person. Recruitment can also be of a person completing an Employment Rehabilitation Course.

Other defined physical handicaps include such defective sight as to be classified partially-sighted or blind, and such poor hearing as to be classified deaf. Impaired vision is immediately apparent to all those with normal sight and evokes sympathy and offers of help. Poor hearing is not so apparent and, at least at first, can cause irritation.

With either disability, an unskilled person is limited, in job opportunities, to work which does not require good sight or hearing. Although both impaired sight or hearing can occur suddenly, with either there may be time, as the condition develops, to arrange retraining. There are specialist societies to turn to for expert advice; several of these arrange training courses if a change of job becomes necessary, and all provide constructive help and guidance.

After an injury or illness most people are able to return to their normal work, but some, especially manual workers, handicapped by a residual physical disability, need lighter work.

Most large hospitals have, or have access to, rehabilitation units providing medical and nursing care, physiotherapy, remedial therapy and occupational therapy. Local authority social services provide a range of personal social services for when the disabled person returns home.

The Disabled Persons (Employment) Acts have been under review for some time. They have the admirable aim, as stated in the relevant government leaflet, of helping all those who wish to work for a living but are handicapped by disability from getting or keeping suitable employment. The leaflet points out that experience has shown most disabled people can take their place in the working community and if the occupation is chosen carefully, hold their own in competition with others. Disabled people may have less sick leave than their non-disabled colleagues.

Since 1974, the Manpower Services Commission has been responsible through the Employment Service Division (ESD) and the Training Service Agency for industrial rehabilitation, vocational training and resettlement in work. This disablement advisory service places many thousands of disabled people in suitable work each year.

The ESD Specialist Disablement Resettlement Officers (DROs) (there is at least one in each employment area), are trained to help disabled people and guide them into suitable work. To qualify for this help the disabled person should be on a Disabled Persons' Register (DRP), maintained by the DRO.

To qualify for the register a person must have a disability which substantially handicaps and is expected to last at least twelve months. Application starts with completion of form DP 17, obtained from Local Government Employment Offices. Medical evidence of disability is usually required, and is supplied by the applicant's GP in a short confidential medical report on the form DP 3, which is provided with form DP 17. This report is not covered by the NHS and the GP is entitled to make a charge.

Queries about eligibility are considered by a panel from a disablement advisory committee, made up of representatives of employers and employees, plus a doctor and those specially interested in disablement. Once approved, the entry is added to the Register: a Certificate of Registration, the 'Green Card' is issued (blue in Northern Ireland). Registration lasts between one and ten years, depending on the nature of the disability. Under certain circumstances, the ESD may have a name removed from the Register and the registered person may remove their own name at any time.

There is (except for merchant seamen) a Code of Practice under the Acts for every employer with twenty or more staff to employ a 'quota' of registered disabled people which is 3 per cent of the total staff. This quota scheme has been the subject of a discussion document.

By regular liaison with employers in their areas, DROs will know when suitable vacancies occur.

Two occupations are reserved for the registered disabled – car park attendant and passenger electric lift attendant. But few lifts nowadays need an attendant and for some disabilities, being in a car park, possibly out of doors in all weathers, and breathing exhaust fumes, is far from ideal

Towards the end of the period of registration, application may be made for renewal. If registration lapses, and even if that person is no longer disabled, he or she still counts towards the employer's quota, so long as they are with the same employer as when put on the register. When the proportion falls below 3 per cent, the employer must consider a registered disabled person for every vacancy that occurs. In practice, this quota system does not work well, and increasingly 'bulk' exemption certificates (DP 42s), valid for six months at a time, are issued by the Manpower Sevices Commission. Possession of a certificate enables a company to employ an able-bodied applicant whilst below the quota. To do this without a DP 42 is an offence. Most firms are below the quota: numbers on the DRP have fallen, the total labour force has decreased, and it is impossible for all firms to maintain a 3 per cent quota.

If a sick employee is likely to have difficulty in returning to his or her previous job because of residual disability, the sooner he or she sees the DRO the better. Too often the DRO is not involved early enough. Mostly, introduction is at the Job Centre where the DRO is based, but there are some Hospital Resettlement Officers (HRO) who visit both in-patients and out-patients at the invitation of a member of the medical staff or a social worker.

At the request of the DRO or doctor, the ESD arranges for a detailed assessment of working capacity at one of the twenty-seven employment rehabilitation centres (ERC); these are mostly nonresidential and were formerly called industrial rehabilitation units. ERCs are usually situated in the grounds of Skill Centres and reproduce working conditions. The role of the ERC is to assess, give vocational guidance and help restore fitness to work and self-confidence. Attendance is for an average of eight weeks. After this the disabled person may go on to a six month training course at a Skill Centre or college.

The Training Service Agency provides training for individuals, including the disabled, who want to learn new skills. There are over 500 training opportunities courses (TOPS).

The DRO may find a 'sheltered' job for a disabled person unfit to work under normal conditions of employment, but capable of useful work. Sheltered work is, for example, provided by Remploy (a nonprofit organisation set up by the government in 1946), by local authorities and by voluntary organisations.

Despite these efforts, a lot of disabled people of working age are unemployed. Since 1979 the Manpower Services Commission awards up to one hundred trophies each year to companies with the most constructive policies on employing the disabled.

Examples of legislation in other EEC member states are as follows:

(1) WEST GERMANY – an employer of sixteen or more people has to employ 6 per cent registered severely disabled (judged by loss of 50 per cent or more earning capacity decided by a specially appointed doctor). An employer with less than this quota pays a

monthly levy to a fund to promote the employment of such people.
(2) FRANCE – there is a 8 per cent quota. Certain jobs are
reserved for the disabled and when one of these becomes vacant
the company must notify the local employment office and appoint a
disabled person who applies within the next fifteen days. The
employer is reimbursed 80 per cent of the cost of any modification
of the workplace and 50 per cent of the cost of special supervision.
(3) ITALY – in general the quota is 15 per cent for employers of
thirty-five or more people. There are complex exceptions; certain
jobs are 50 per cent reserved for the disabled.
(4) BELGIUM – reduction of working capacity by 30 per cent or
more from a physical cause or 20 per cent or more from a mental
cause, qualifies as disablement. There is a national fund for placing
these people and providing subsidies to employers. In the public
sector there are reserved jobs.

There are six centres in this country with permanent exhibitions of
aids for the handicapped. An example in London is the Disabled
Living Foundation Aids Centre. Funded partly by the Department of
Health and partly by charity organisations, the centre is open to
visitors by appointment and gives the opportunity to inspect most of
the aids available for the disabled. These range from wheelchairs to
ear-level flexible telephone handpiece holders. The Disabled Living
Foundation publishes booklets of practical advice for the disabled;
videotapes and films are also produced.

Apart from being involved in the design and siting of health
services units, company doctors and nurses can usefully advise ar-
chitects of new office developments on the provision of facilities for
the disabled. All new and converted office buildings should cater for
such people and all buildings open to the public must now do so.
Under the Chronically Sick and Disabled Persons Act, clearly sign-
posted access and toilet facilities for disabled people should be
provided.

Detailed guidance is given in publications of the Disabled Living
Foundation, including dimensions of the size and turning circle of
wheelchairs. Dimensions of lifts, doors and toilets have to be designed to accommodate and allow manoeuvre of wheelchairs. In the
toilet, the height of the lavatory seat, wash basin, taps, grab rails and
coat hooks are apt to be put in standard positions that are too high for
a person seated in a wheelchair to reach.

Consideration must be given as to how the disabled person would
escape from the building in the event of an emergency such as a fire.

A priority lift should be designated to evacuate disabled persons as soon as the alarm is sounded. In a high rise building, this system can be clogged by people with nothing much wrong with them who simply don't want to walk down many flights of stairs.

The hospital occupational therapist advises on site about aids for the disabled: door handles, doorway widths, switch positions, ramps and rails, for example. There must be close co-operation with representatives of the company's premises staff who will know what is and what is not reasonably practicable.

Severe disability

There is an attendance allowance for those so handicapped as to require prolonged or frequent nursing attention. The invalid care allowance is for those of working age who have to stay at home to care for a disabled relative receiving an attendance allowance. There is also a mobility allowance for those who have difficulty in walking.

An example of severe disability is where there has been damage to the spinal cord, as can occur from injury in a fall, or in a traffic or sporting accident. This damage is permanent, and may leave the often young person with a normal brain, but paralysed legs (paraplegia) or paralysed legs and arms (tetraplegia) and confined thereafter to a wheelchair. Both movement and feeling may be affected and, if the injury was to the brain, epilepsy is a possible sequel.

Numbers of paraplegics and tetraplegics have increased significantly in recent years both because of the increased incidence of such accidents and because advances in treatment prevent death soon after the accident.

Paraplegics and tetraplegics can have the problem of urinary incontinence. This is contained by the wearing of a disposable bag attached to the leg, or by a highly absorbent pad contained within leakproof pants. Once again the only requirement at work is a nearby toilet, adapted for use by someone in a wheelchair with partial or total paralysis of limbs.

PERIODIC HEALTH SCREENING

In occupational medicine there has been a tradition of health screening since the Factory Acts of the mid-nineteenth century, when doctors were appointed to certify the minimum age and later, fitness, of children working in factories.

Regular checks are statutorily made in certain industries, of employees working for example in the environment of lead fumes or mercury. Before including a particular test, the following criteria should be fulfilled:

(1) The condition sought should be an important problem, with a latent or early pre-symptomatic state.
(2) There should be an acceptable test and treatment for the condition and an agreed policy of whom to treat.
(3) Facilities for diagnosis and treatment should be available.
(4) The natural history of the condition should be adequately understood.
(5) Screening should be a continuing process, not 'once and for all'.

Screening is popular. 'In-house' screening by the company doctor or nurse who know about the work practices is one good way to carry this out, with it clearly understood that, statistics apart, screening is confidential, with no reporting back to management. But there may still be, without foundation, the suspicion that the doctor or nurse will inform the personnel manager. This just would not happen. For companies without a screening facility, there are plenty of organisations ready and able to take on this work; one advantage of outside screening is that they are clearly nothing to do with the company. One screening package includes a self-administered health questionnaire, filled in before attendance, on mental and physical health and life style. The number of questions is gauged to avoid boredom or mental fatigue. Answers can be probed later, when there is the opportunity for consultation with a doctor who, for the sake of continuity, should be the same doctor on each visit, so that a relationship can be built up. The same doctor should see all the people from the same company.

The measurements and tests include measuring height and weight and resting heart-rate and blood pressure; an electrocardiograph; urine analysis; the X-ray of chest and abdomen, and tests of hearing, eyesight and lung function. A blood sample is taken for various analyses such as blood fats. The results of the blood tests may not be ready for a day; this is a disadvantage covered by the examining doctor writing later to the person about their normality or otherwise.

The doctor's summary, together with the results of all tests, are sent to the person's GP who decides on their relevance and the need for further investigation or treatment. There is the risk of the screening doctor making one recommendation and the GP another. In some countries, and increasingly in the UK, the person who has been screened gets all the detailed results, ECG report and so on, allowing him or her to see how these compare with the norm. Where the cost is paid by the employer, and with the written consent of the employee, a copy of the report may be sent to the company doctor or a non-technical summary sent to the personnel manager.

Some companies hold the view that reporting-back to personnel would be counter-productive in as much as it may either discourage staff from having the medical, or if staff think that to refuse the medical could make them conspicuous, lead to their being less than totally honest, and so little benefit would result. These companies rely on the individual telling the personnel manager of any findings that might affect their work.

Screening is not infallible but some people think that it is, and believe that regular annual or biennial comprehensive medical examinations are necessary to ensure continuing good health – that just as a car or any piece of machinery needs regular servicing, so do human beings. Continuing good health cannot be guaranteed, nor can all developing disorders be detected, by a battery of tests, however impressive they appear to be. It is important that a company does not use a normal health screen as justification for increasing an individual's workload.

Critics of general health-screening of healthy adults point out that what happens in between screenings matters just as much as the state of health on the day of the screening. Screening may create needless anxiety when something is found which either does not matter or about which nothing can be done anyway. Life-style health advice on weight, exercise, diet, smoking, and so on, could often be given without any tests. Most of this advice should be obvious in any case to an intelligent adult. Comprehensive medical screening is expensive in time, medical staff and money. Statistically, comprehensive medical screening does not prolong or improve the quality of life. Most screening for heart disease does not start until middle age; any screening programme should ideally commence early in life when life-style advice, if acted on, could be more effective in the long term.

On the other hand, those in favour of screening point to the many silent disorders which are curable or treatable if identified early: for example, to detect, investigate and treat high blood pressure is life-saving. Also, people are more likely to take notice of and act on advice based on what their own test results reveal.

To illustrate the fact that health screening cannot guarantee continuing good health, the cynical story is told of a doctor who over the years built up his practice solely on screening executives. Soon after the doctor had completed one such examination, reassured the male executive of his good condition, and said goodbye 'until the next time', the doctor's receptionist rushes in saying the man has dropped dead outside the doctor's front door. 'My reputation will be ruined', exclaims the doctor, who then has an idea: 'Turn him round,' he orders, 'so it will look as if he was on his way in for his screening.'

In some companies, screening examinations are only available to senior management, who are costly to employ and replace, have much vital decision-making stress, whose lengthy absence through illness would be catastrophic, and to whom the company wants to show 'it cares'. So eligibility for a screening medical is for some a status symbol or perk. It could be argued that senior executives have learned how to cope with (and even thrive on) high stress levels and so may be at less risk of stress-related illnesses than those below them in middle management grades who are aspiring to greatness, but not necessarily reaching it. The bored, frustrated clerk also runs this risk. There is no hard evidence that as a group, senior management are at more risk of illness than anyone else of their age. Most executives in the company who are of the status which qualifies them for screening, are middle-aged, and changes in the body from wrong eating habits and other aspects of an unhealthy life style not only are already present, but having developed over many years, may not be readily reversible. Also, the middle-aged person is more set in his or her ways, and if feeling well is perhaps less likely to heed advice to make changes in life style.

The good news, though, is that coronary heart disease from atheroma (narrowing of the coronary arteries, often the cause of a heart attack) can be reversed, as repeat coronary X-ray angiograms of the coronary arteries show some years after making changes in life style. One American, Nathan Pritikin, with this proven diagnosis at thirty-nine promoted a low-fat diet and vigorous exercise regime in lecture tours around the world for thirty years, then died of an unrelated condition. The post-mortem revealed arteries as smooth, pliable and soft as those of a child.

The 'cut-off' point of availability of screening at a certain level of management can cause resentment in the main work force. An alternative is a simpler but no less meaningful screening programme available to all staff.

The philosophy of health screening is summed up on the cover of the health leaflet which is given to all new National Westminster Bank staff: 'much of the responsibility for achieving and maintaining health and harmony of mind, body and spirit, i.e. wholeness, rests with each individual and that person's attitude to his or her total life style'.

A screening programme which could be carried out by the company nurse could include the following:

Basic screening tests	*Purpose of test*
Short, confidential questionnaire on aspects of health including smoking and drinking habits and perception of stress.	Assess life style. Detect early symptoms of disease.
Pulse rate at rest	Detect abnormal heart rate and rhythm.
Blood pressure at rest	Detect high blood pressure.
ECG on first visit (optional)	Base line reading.
Lung function; peak flow rate	Demonstrates asthma, lung fibrosis and adverse effects of smoking.
Height and weight	Gives body mass index of obesity (weight in kg divided by height in metres squared should be 20 to 25).
Urine analysis	Detect diabetes, kidney and other urinary-tract diseases.
(Stool test for blood – but may give false positive)	Detect cancer of bowel
Blood test if family history of heart attack under age 50, or signs of cholesterol deposits	Detect raised blood–cholesterol
Eyesight, visual acuity and visual fields	Detect abnormality of vision

Literature provided when indicated for:

Exercise
Losing weight by healthy eating
For women, early breast-cancer detection by self-examination

Food fat finder
Advice on coping skills

All of these tests can be carried out by the company nurse. Abnormal results are referred to the GP.

Frequency of the examination depends on age, sickness record and results of previous tests; every five years would be suitable for the healthy person. A person with 'borderline' blood pressure should have this checked every six months.

A high blood–cholesterol level is a risk factor for heart disease. This may be an inherited trait with a history of a heart attack in a close male or particularly female relative below the age of sixty. With TB no longer prevalent in the UK there is no need for a routine chest X-ray.

A regular screening chest X-ray can even encourage a person who smokes to continue smoking, as this X-ray principally looks for evidence of lung cancer, and a smoker whose X-ray is clear, may take this as a licence to continue smoking, at least until the next X-ray. But a chest X-ray is unlikely to detect very early lung cancer and the cancer will have a firm grip by the time it shows up. Immediate treatment of a symptomless cancer detected by annual chest X-ray does give better results than when treatment is delayed until symptoms occur, but the outlook is still grim.

A normal resting electrocardiograph (ECG) means to the lay person that the heart is normal. In fact, whilst an abnormal ECG may be significant, a normal resting ECG does not exclude heart disease. The ECG is a far from perfect guide to heart function and is more an indication of what has happened in the past than a predictor of what will happen in the future.

However, a routine ECG can provide a base line for comparison later, if the need arises, with an ECG being taken because of, say, an episode of chest pain. Any change in the pattern of the tracing will be significant, and confusion with a long-standing abnormality, because it has been previously recorded, avoided. The significance of a minor change can be difficult to assess, and knowing it is there may make that person worry unnecessarily.

A resting ECG shows changes in heart-rate and rhythm, provided these occur during the test; disorders of conduction; enlargement of part of the heart or damage to an area of heart muscle, and the effects of certain drugs on the heart.

If symptoms suggest the possibility of short bursts of abnormal heart-rate or rhythm, a twenty-four-hour or longer ECG is recorded on tape during normal activities on a special miniature casette-recorder. The tape is then analysed by a computer which detects any variation from normal.

Some abnormalities fail to show up on the standard 'resting' ECG. For example, about half of those with angina and significant heart-muscle damage, have a normal ECG at rest, with these abnormalities only showing up on an ECG recorded during and immediately after a standard controlled exercise programme – the so-called 'stress' ECG which is only carried out with resuscitation equipment on hand because of the possible risk of precipitating a heart attack. Once more there is the risk of a false positive, when there would be the need for further tests to prove normality and in the meantime causing needless worry.

One example of a silent condition which can be effectively treated if detected early is glaucoma. The pressure is the eye gradually rises and reduces the blood supply to the retina; the outer area of the retina is damaged first; this affects peripheral vision and may be unnoticed. The condition worsens, but central-focusing vision remains intact until late on, and those affected may not realise their eyesight is restricted to 'tunnel' vision until they start to bump into objects. By then considerable irreparable damage to the retina has taken place. Incidence increases from age forty.

Glaucoma can be detected before any sight is lost by measuring eye pressure (tonemetry) with a flattened, cylindrical lens against an anaesthetised eyeball. An alternative system uses a puff of air. Anyone over forty should be tested every two years; many opticians now have this air test. If glaucoma runs in the family, the test should start even earlier.

WOMEN'S HEALTH SCREENING

Cancer of the cervix, the neck of the womb, kills 2000 women every year in the UK. Screening for pre-cancerous changes, the cervical smear, is presently available under the NHS every five years to women over thirty-five, or to women who have had three or more pregnancies. There are NHS proposals to provide the test every three years for women over twenty. The test itself is very quick and only takes a few minutes. Cells are painlessly removed from the surface of the cervix with a special spatula or brush. The cells are smeared (hence the title) on to a glass slide, fixed and sent to a laboratory for microscopic examination.

Proven abnormality is treated by local surgical removal or laser beam long before abnormal cells become cancerous and spread into or beyond the neck of the womb.

To my mind, providing this service at work duplicates what is now an improved NHS facility and overloads stretched pathology facilities. Better to encourage women, who might not otherwise have done so, to attend their own GP, giving them time off to attend. The test can be performed by a technician, a nurse or a doctor; with the latter two there is also the opportunity for discussion of any gynaecological symptom; the doctor also performs a pelvic examination. The Women's National Cancer Control Campaign provide a lady doctor and nurse team, and for a set fee see twenty-five women in three hours, using either a mobile unit parked outside or suitable rooms on the company's premises. A company employee makes the appointments, acts as receptionist, and fills in the forms.

Sometimes also checked during a women's health screen are weight, blood pressure, and urine. Ensuring knowledge of breast self-examination is one of the other important procedures, specific to well-women screening and concerned with the fact that each year in Britain over 40 000 women are treated for breast cancer and each year 18 000 women die from this disease – a rate of twenty-eight per 1000 women. Breast cancer is the commonest cancer in women and strikes one in twelve women in the UK; this is the highest incidence in the world.

Risk factors include a family history of breast cancer in either sister or mother below age fifty, which increases the risks frighteningly to between one in three and one in four; periods starting at an early age; a late menopause and, curiously, a high fat diet. Most breast lumps turn out to be benign and innocent, but all need investigation, involving one or more of the following: clinical examination, mammography, ultra-sound or aspiration. These investigations can be carried out at a breast clinic, such as the Early Diagnostic Centre at the Royal Marsden Hospital, London.

There are several different kinds of breast cancer. The vital factor is how often cells double in number. If this is infrequent, the cancer is likely to stay localised for years. Where this time is short, the cancer quickly spreads and it is the growth of this secondary spread away from the breast in other sites which kills. The best chance a woman has is the earliest possible detection of the cancer by regular screening.

A simple method is for a trained nurse to demonstrate and teach routine, thorough, monthly self-examination.

A woman may not act on finding a lump, being too fearful or embarrassed to do anything. Another screening method finding favour is regular mammography – very low-dose X-rays of the breasts which show up early cancer change before it can be felt. At one centre, twenty-five per cent of breast cancers found by mammography over an eight year period could not be felt. Mammography is most effective in women over age 50 when the breast tissue is less dense and breast cancer becomes more likely. Mammography techniques now use lower dosages of radiation, well within recommended safety limits. Private schemes recommend an earlier start, the first test is at age thirty-five and then every two years from forty to fifty, then every year over fifty; with a positive family history, testing starts at twenty.

Mammography is a picture at one point in time and between tests an 'aggressive' cancer can develop, so this test, which has been studied by the NHS and shown to be superior to anything else, should be combined with monthly self-examination.

Ovary cancer presents late, but clinical examination, plus ultrasound, plus a blood test, can detect this before symptoms develop.

CANCER

General Advice

Cancer describes a group of diseases which can often be cured, especially with early treatment.

The following symptoms do not necessarily mean cancer but should be discussed with your G. P.:

A sore that does not heal.

Enlargement, change of colour or bleeding of a wart or mole.

A painless lump.

A *persistent* cough or hoarseness.

Difficulty in swallowing.

Persistent indigestion.

the work stations in the office during working hours. More often a space is cleared in a rest room or staff restaurant immediately after work for a more strenuous aerobics class – a continuous exercise routine to music which goes on for up to an hour at a time. It was mostly women who attended these classes – men are now joining in. The class is run by a leader: better a trained professional who charges a fee rather than an enthusiastic amateur who may be more concerned about his or her own ability than supervising others. A badly-run class does more harm than good. To experience 'the burn', as the muscle pain is called, is overdoing it. Changing facilities should be available, with showers. I prefer a short set of timed and supervised exercises to aerobics: thirty seconds exercise – thirty seconds rest to record the pulse rate and move to the next – twenty in all.

A few companies have their own swimming pool where it is essential to have life guards trained in rescue and resuscitation permanently on duty when the pool is open.

Likewise a few companies have a gymnasium and even expect all staff to go there at least once a week. A cynical view is that once the novelty wears off, the gym will only be used by those who would keep fit anyway, and there is the experience of one company where the company paid a limited number of memberships for a new health club, and those who had free membership never actually turned up – maybe feeling fitter from merely belonging! The apparatus in a gym may include a computerised stationary bicycle for fitness assessment, rowing machine, powered treadmill, weights and machines for shifting weights, parallel bars and wall bars. Clean, well-ventilated changing facilities including showers are essential. A compact, practical gym takes up very little space.

A gym should be designed and supervised by a trained instructor. Everyone needs an individually-tailored programme, devised and supervised by an expert; guidance cards prominently displayed remind what that programme is.

Medical clearance, in the form of a consultation with the GP, is prudent before starting a fitness programme for a person who has a history of significant ill health, especially of a heart condition. The GP should also be consulted if any significant symptom develops during the fitness programme. Of course, some breathlessness during exercise is to be expected, but not to the point of gasping for air or exhaustion. It should be possible to talk at the same time as taking exercise. More objectively, the pulse rate per minute taken at the wrist should rise to about three-quarters of 220 less the person's age during exercise and return to below 100 within two minutes of stopping exercise.

A company gym may be used for rehabilitation after a heart attack, in which case expert supervision in liaison with the individual's doctors is essential.

Criteria used in determining an individual's training programme include age and degree of lack of fitness, the latter being gauged to an extent from the pulse rate. Any training programme should start gently and gradually build up to incorporate more strenuous routines. Don't exercise after eating or if you feel ill. Jogging has the advantage of being simple, needing no special skills – an uncomplicated way to get fit. Injuries can occur to knees, shins, ankles, heels. Achilles tendons and occasionally hips, not to mention the obvious blisters. Then there is 'jogger's nipple' – a dermatitis caused by friction and prevented by two pieces of sticking plaster.

Some of these misfortunes happen to those who do too much too soon, so train – don't strain. Wear good quality, lightweight shoes with cushion soles. Limber up and stretch and try to jog on soft ground, not roads or pavements.

Anyone with chronic joint disorders should take special care, also the overweight and the middle-aged, who are more at risk of ligament strain and back injury. On a positive note, an osteopath said that jogging can help bad backs, as during a twenty-minute jog the spine has 2500 gentle manipulations which loosen joints and strengthen supporting muscles.

Swimming is a more gentle and non-weight-bearing activity and does not generally produce the health problems – muscle and joint injury – of ground contact sports. After a few weeks on a fitness programme, there is enhanced muscle strength, suppleness and stamina.

Competitive sport must be 'worked up to', and should always be against someone of the same age group.

The 'warm-up' and 'warm-down' referred to consist of stretching limbering exercises which lessen the chances of stiff, strained muscles and ligaments.

It is never too late to start an exercise programme, but always begin gently and build up. It was said that everyone should first check with their doctors, but whilst this is true for someone who has, for example, angina, usually it should not be necessary. It is rather those who decide *not* to take exercise who should be consulting their doctors!

HEALTH OF DRIVERS

Sales representatives and chauffeurs are examples of professional drivers who, during each working day, can be in their cars for many hours. Running late for an appointment, frustrating traffic jams, noise, fumes and vibration, are all sources of stress, plus the need to stay alert in all weather conditions for long periods at a time.

Although the collapse of a driver at the wheel is rarely responsible for an accident, with most accidents due to risk-taking (more common in smokers who drive, who are knowingly taking risks with their health by smoking and apparently are inclined also to take risks when driving), adverse road conditions, alcohol, drugs or sheer incompetence, there is a case for all professional drivers to have a periodic medical examination at intervals related to age and health record. This is partly in line with an EEC directive which proposed psychological tests and a medical examination as well as a driving test before the issue of a licence, with periodic medical examinations thereafter at decreasing intervals as the licence-holder gets older.

In the United Kingdom, a driving licence for a private car is valid to age seventy. An applicant for a driving licence must declare whether or not he or she has epilepsy, suffers fom sudden giddiness or faintness, has eyesight less good than the required standard, or has a condition, for example multiple sclerosis, which could become a relevant disability – this is called a prospective disability. A declaration of continuing good health is no longer necessary every three years, but by law a driver must inform the licensing centre at Swansea of any relevant or prospective disability as soon as possible or when a declared disability becomes worse. (Only a temporary disability like a fracture which is not expected to last more than three months is excluded from this obligation.) The medical adviser at the Licensing Centre then becomes involved and may seek a report from the GP concerned and/or arrange a medical examination. The decision may be reached that the driving licence has to be surrendered.

The eyesight standard includes the ability to read a car number plate, that is, letters and numbers three and a half inches high, from twenty-five yards away. A licence may be granted for up to three years to a person subject to epilepsy, who has been free from any attacks, with or without treatment, for at least two years, or who has had a three year pattern of being subject to attacks only whilst asleep. In all cases the criterion is that the person will not be a source of danger to the public.

Drivers of heavy goods (HGV), public service vehicles (PSV), and forklift trucks, are the only drivers who are required by law to pass a medical first, before issue of the licence, and periodically thereafter. For HGV drivers this means renewals from age sixty and for a PSV driver, renewals from age forty-six, and annually from age sixty-five. For drivers of these vehicles, health standards are higher than for the driver of a private car. For example, a history of angina or heart attack must be fully investigated and may preclude further HGV or PSV work, as does insulin-dependent diabetes or a single epileptic attack after the age of five. A regular health screen examination of an essentially healthy, symptomless driver includes testing eyesight and fields of vision to ensure the absence of a 'blind spot', hearing, pulse rate and rhythm, blood pressure, height and weight, limb function and ensuring freedom from a disabling or restricting neck or back disorder.

Amongst the twenty-five million UK drivers, sudden illness accounts for about one accident in a thousand, and it is more a combination of what could even be termed a personality disorder with disturbed behaviour, aggression or indecision which is responsible for most accidents.

The health screening looks in particular for heart disease, which could cause sudden collapse at the wheel and injury to others (three-quarters of such collapses are from this cause). A history of a heart attack or angina rules out professional driving, as do most abnormal results of tests on the heart like an exercise electrocardiagraph or coronary angiogram – a test which shows the blood supply to the heart itself. No one should drive within two months of a heart attack; after a successful coronary artery bypass graft, a person may return to driving on recovery from the operation.

Although more accidents occur to young drivers, with the highest incidence between the ages of twenty and twenty-three and in the first two years after passing the driving test, there is merit in offering the screening examination both before appointment as a company driver and periodically thereafter, say at ages fifty, fifty-six, fifty-nine, sixty-two, sixty-four and then annually until retirement.

The sickness absence record of each driver should be reviewed every month, with any relevant absences such as those due to giddiness (vertigo), heart or circulation disorder, epilepsy, eye disorder, diabetes, backache, or for longer than three weeks, referred to the company doctor.

There are legally-defined upper limits for time spent at the wheel and distance covered each day by HGV and PSV drivers, but not for chauffeurs, van drivers or sales representatives.

Chauffeurs may spend a lot of time waiting, sometimes in a smoky waiting room or in the car. By late evening, when the time finally comes to drive the executive home, the chauffeur may be tired and so at increased risk of an accident. Because of this the executive and chauffeur may share the driving on a long or late journey.

The sales representative who has been driving on and off all day long trying to make appointments on time, is stressed by the frustration of a traffic jam while running late for an appointment, and with no way of letting the customer know. This stress could be eased by installation of a car telephone (although of course this should not be operated when the car is moving.)

When the wearing of seat belts by the driver of the car and the front-seat passenger became compulsory in the summer of 1982 under the Transport Act 1981, doctors were provided with Certificates of Exemption to supply when there is a medical reason for exemption. As the wearing of seat belts reduces both fatalities and all types of injuries by about 50 per cent, with reduction in injury particularly applying to injuries to the head and face, these certificates should only be issued in very special circumstances. Modern inertia-reel belts, correctly applied, can be worn with most disabilities and in pregnancy.

Back disorders in drivers are not uncommon and may be related to the fact that some drivers seats do not provide adequate low-back support. Sometimes the shape of the seat is wrong, but often, as in the office, it is because the seat is not adjusted properly.

It saves time (but causes strain) to twist and lean over from the front seat to the back of the vehicle to collect items; it is difficult to keep a straight back when lifting suitcases and other items in and out of the boot, especially if there is a high sill.

Instruction in the right way to lift is as important for drivers as for other staff.

HEALTH OF CATERING STAFF

Around three million people work in the catering and retail food industries. Some companies with large numbers of staff concentrated in a few locations provide 'in-house' catering, using either their own catering staff or contract caterers. One advantage of employing one's own staff is being able to ensure good health surveillance – one disadvantage is that, as a generalisation, catering staff have more sickness absence than clerical staff and there may be the need for temporary agency staff, whose health cannot be so closely checked, to provide cover.

All food handlers should have a health interview or medical examination before first starting work. This will look for any condition such as recurrent skin infection which may harbour germs like staphylococci. Some people should not work with food because of increased risk of food poisoning, particularly anyone who is subject to recurrent sore throats or discharging ears or nose, or boils or diarrhoea. Whilst sickness absence may be higher in catering staff than other employees, there is no real reason why this should be so, except that any suspicion of a condition which could cause food poisoning in others stops that person handling food. Principally concerned are skin infections and gastroenteritis, which would not necessarily prevent clerical workers from coming into work.

The food-handler should be in good general health and not a persistent 'silent carrier' excretor of illness spread by food, like typhoid, paratyphoid or hepatitis A; at the turn of the century, 'Typhoid Mary' was a typical example: she poisoned every household she worked for. The catch question put to medical students is 'Was she a good cook?' The answer is 'Yes – why else would other households have taken her on?!' A food-handler should also maintain high standards of personal and dental hygiene and not be a nail-biter. In the confined, high-accident-risk environment of a busy kitchen, a handicapped, disabled person may be a hazard, so this is one area from which people with some disabilities should be excluded. A clear chest X-ray was a common requirement for food-handlers, to exclude tuberculosis of the lungs; but as human tuberculosis is not transmitted through food, X-ray for tuberculosis served no more than to protect other catering staff in the same unit. Tuberculosis has become less common in the UK so this requirement has been discontinued, but it should be carried out overseas, where tuberculosis is rife, on any local staff to be employed in an expatriate's household.

One way to lessen the risk of food poisoning is to require a negative stool test from all catering staff. This is a microbiological check for bacteria unwittingly harboured by a carrier, which, given certain conditions, could cause food poisoning. This is standard practice in many food packaging plants but only in a few company catering units. The snag is that routine stool tests to detect symptom-less carriers of food-poisoning germs are fine in theory, but in practice excretion of these germs may be intermittent and so may be missed however many samples are checked – one negative is not enough, three should be, but are no guarantee, and it can be difficult to get the co-operation of a person who is feeling quite well. This test could make sense, though, for existing staff, if after a holiday over-seas, the catering worker discloses a history of enteritis whilst away. The same applies during or on return from short-term sickness absence, when a GP may not have been consulted. The condition could be used as reason to stop work by a food-handler who is *plumbi oscillans* – swinging the lead – who finds that a diarrhoea illness means an immediate ban on kitchen work. A company doctor or nurse can prove the diagnosis quite quickly by providing a specimen pot and saying 'return within the hour with a sample.' To be practi-cal, provided the food handler who was ill has recovered and has a solid stool and good hygiene, the possibility of him or her causing food poisoning is remote.

Understandably, a catering manager may look on a diagnosis of food poisoning at work as a criticism and a slur. The chain of food poisoning can start with bacteria from a food-handler getting on to suitable food which is being prepared. All staff handling, cooking and serving food must comply with hygiene regulations in the Catering Acts, the OSRP Act, and all other acts affecting catering establish-ments. Active infection in a food-handler may cause food poisoning in others, especially if the food-handler has a gut infection causing diarrhoea, septic discharge from the ears, eyes, nose, mouth or throat, or septic skin lesions such as boils anywhere on the body and not just on the hands or face. Should food-handlers get one of these conditions, the GP must be made aware of the nature of their work.

A food-handler must keep clean and have good personal hygiene – not easy in the heat of a kitchen. Providing a shower in the changing rooms helps, but is not always used in practice. Finger nails should be kept short and clean but not bitten. Have reasonably short hair or keep long hair covered. Wear suitable footwear – non-slip and with protected, reinforced toe caps including over the little toes. Wear no jewellery (except a wedding ring). Cover a cut or burn which, however small, can harbour bacteria. By law – Food Hygiene (Gen-eral) Regulations 1970 – any cut or burn must be covered by a *blue* waterproof plaster (blue to spot readily if by chance the plaster falls into food). Stop work if suffering from a rash or boil, particularly on the hands or arms, or from a discharging ear, a sore throat, sickness and/or diarrhoea. Do not smoke whilst working.

Kitchens are high-accident-risk areas so it is common sense to ban distracting personal stereos.

As mentioned earlier, diarrhoea in a catering worker rules out kitchen work – management may on occasions suspect the symptom is invented to avoid work. Much sickness absence is self-certificated; the GP is not necessarily consulted and need no longer provide a 'freedom from infection' certificate. In all cases of acute diarrhoea, however, the employee must be urged to consult his or her GP, if necessary taking a letter from the company pointing out that the employee is a food-handler.

The GP may arrange a stool test and check it is negative before passing that person fit to resume work. If bacteria such as salmonellae are found, three consecutive negative stools are required to prove clearance; but even then excretion of these germs may be intermittent as previously mentioned.

Staff working with contaminated raw foods may by chance take in the germ and excrete it so that excretors of salmonellae may be victims of the foods they handle.

A doctor is required to notify a food-poisoning illness to the local authority, and under the Food and Drugs Act, the employee who is aware that he or she is suffering from, or is a carrier of, any infectious agent likely to cause food poisoning, must notify the employer. The employer must also notify the local authority, which then carries out an investigation, looking especially at hygiene.

The Food Hygiene Regulations list diseases which must be declared by the employee. The employer's duty in monitoring this is less clear.

Of equal importance as the health and hygiene of catering staff, is the hygiene of the working environment in the kitchen. As much equipment as possible should be made of stainless steel and a detailed cleaning programme must be established. Wherever possible, equipment should be on castors, so that it can be moved to allow cleaning behind it. To prevent the spread of bacteria from raw to cooked food, kitchen cloths should all be disposable.

Care of catering equipment in branches and departments is also very important and tends to be overlooked. A refrigerator is an obvious example. There may be no one responsible for regular, routine defrosting, or cleaning of the inside and of door seals. Someone should be responsible for disposing of out-of-date items which otherwise may be kept in the refrigerator for weeks. Where a drinks vending machine is in use, someone specially trained should be responsible for not only the refilling, cleaning the outside and restocking with disposable cups, but also for regularly – that is, daily – flushing and sterilising the tubing through which the water flows. Regular microbiological checks should also be carried out.

The same scrupulous care must be taken when food is prepared by the 'cook rapid chill' method, stored refrigerated for a short shelflife and then regenerated. Likewise when cooking 'sous vide' in a vacuum.

The concern about the finding of salmonella bacteria in poultry and in raw eggs emphasises the need for thorough cooking.

The use of microwave oven in staff rooms is quite safe provided the equipment is properly used and maintained.

8 Overseas Travel

THOSE AT RISK

Many companies have overseas interests, with expatriate staff working abroad for two to three years at a time and other staff making short tours lasting from a few days to a few weeks: It is prudent for the company to ensure as far as possible the continuing good health of their staff whilst abroad, particularly for the latter group, who are likely to be moving from place to place when even a minor illness, alone, abroad, is alarming, inconvenient, saps self-confidence and is expensive to the company in lost time.

A tropical climate may contribute to the worsening or relapse of a number of conditions, so a history of one of these is sufficient reason to advise against a posting to the tropics. It is important to arrange the medical soon after selection for an overseas job to determine whether the job holder or a member of the accompanying family is subject to any of these conditions. Whilst the job holder and spouse should be seen by the company doctor, it is better that children be reported on by the GP who has the records, details of vaccinations etc. If the medicals are carried out close to the date of departure, because earlier would be inconvenient, or the medical is regarded as a mere formality, or because the company did not want to be committed earlier, arrangements are likely to have been made for letting the family home, changing educational arrangements and so on. To cancel the tour on medical advice at this late stage will be all the more difficult, disappointing and frustrating.

There needs to be preparation for working and living abroad, with information on the way of life, customs, climate, suitable clothing, and so on. There are residential courses for this which cover Africa, the Middle East, South-East Asia, Latin America and the Caribbean. Course members most want to know about security, medical care, housing, housekeeping and schools.

The location within a country is as important as the climate. To live and work in a capital city with good medical facilities is very different from being based in a remote part of that country.

If one member of the family should not go abroad for medical reasons, no one goes. It is no use a fit husband trying to work with the worry of his wife or child sick or at risk from living in the tropics.

100

Nor, if at all possible, should a husband take up the job abroad as planned whilst the rest of the family stays in the United Kingdom. In my experience, splitting the family for any length of time is best avoided.

Any risks to health from living in a particular country must be openly discussed. The sorts of medical conditions for which people with such conditions are advised against working overseas include the following (more strictly applied when the appointment is to a remote area far from medical aid):

> Mental illness, or a history of significant or recurrent mental illness involving hospital in-patient care; out-patient treatment by a psychiatrist for a vulnerable, neurotic personality (additional stresses of living abroad include isolation and, for the wife especially, boredom far from home, family, friends and familiar surroundings. Also there may be hostility from local people or other expatriates.) A mental illness from being abroad usually happens on the first tour, if at all; recent treatment for cancer (follow-up assessment is best in the hands of those who provided that treatment); a tendency to severe diarrhoea such as ulcerative colitis (a relapse can be brought on by a bowel infection, which is almost inevitable in the tropics); peptic ulcer, that is, active ulceration of the stomach or duodenum, or even simply a history of peptic ulcer, especially where bleeding has occurred; tuberculosis; cortisone (steroid) treatment; poorly-controlled epilepsy; multiple sclerosis; asthma (some varieties of asthma do badly in dry heat, others in humid conditions); kidney stones (dehydration encourages stone formation); unstable diabetes; thyroid disorders; gross obesity (as there is poor heat tolerance); severe acne and other infected skin conditions (the skin 'goes bad' quickly in humid heat and there is the likelihood of scarring, particularly worrying and embarrassing to a woman); chronic or recurrent discharging ears.

A periodic health check every three to five years for staff who make short trips abroad for the company is advisable. This neither takes long nor involves complicated tests and can be carried out when vaccination status is assessed and any necessary vaccination boosters given. The interval between check-ups varies with age and health record. For a healthy person every five years is reasonable.

For such short term travellers, the company doctor should be notified of any illness requiring sick leave for more than three weeks and the member of staff told to report any symptom of concern.

It is important that eyesight is checked every two years by an optician, and for the traveller to take abroad the prescription for existing glasses, in case of loss or breakage.

If the date of the last dental examination is more than six months ago, a reminder is given to have a further inspection. It is sensible to ask the dentist to complete, for a fee and from existing records, a chart showing the state of the teeth and what treatment was needed, and confirming dental fitness. A dentist overseas may claim the need for extensive bridge and crown work, which, from reference to these records, is not strictly necessary by UK standards (for example, what is cosmetically desirable may not be dentally essential).

VACCINATION REQUIREMENTS

The vaccination requirements of a country can suddenly change with little or no warning because of an outbreak of a particular illness, preventable by vaccination, in that country or elsewhere. Changes in requirements can be an over-reaction, the country making a requirement for vaccination to prevent import or spread of that disease by the traveller, rather than out of concern for the traveller. So the experienced traveller who may be called to go anywhere in the world at short notice, often prefers to keep up to date with all vaccinations, and, concerned about sudden quarantine, or airport vaccination, takes some dissuading that a vaccination is unnecessary. There is increasing risk of reactions from often-repeated doses of the same vaccine. Repeat vaccinations are immediately effective and in-date no matter how long since the previous dose (but the traveller must continue to carry the earlier certificates as proof of previous vaccination). One snag from vaccination just before a trip is of side effects spoiling that trip, as once happened apparently to members of a band, reacting to injections in the arm and preventing their concert taking place. Many a holiday has been spoiled by reaction to last-minute typhoid vaccination going on for several days.

Sudden changes add to the confusion about vaccination requirements, and that is the reason why detail is given here of available vaccines and, in broad terms, where they are needed. The embassy of the country concerned should provide up-to-date requirements, but their airport immigration authorities may give different information.

WHO produce an annual handbook, available from HMSO with further supplements during the year, which shows at a glance the mandatory vaccinations for most countries.

The 'Travel Information Manual' is a monthly publication by fifteen IATA member airlines that gives current information about mandatory and recommended vaccinations and the malaria risk for every country. The manual also contains general health advice and information about passport, visa and permit requirements, customs regulations, and so on. It is available on subscription from PO Box 7627, 118 Zj Schipol Airport, The Netherlands.

Another source of authoritative information is the WHO weekly epidemiological bulletin. A telex summary of information of importance on all communicable diseases with respect to international travel is prepared for transmission each Friday.

An alternative is to subscribe to the Medical Advisory Service for Travellers Going Abroad (MASTA), which provides daily updating on all health aspects of travel to a particular country, using a VDU screen, keyboard and printer linked via an independent telephone line to a master computer. British Airways vaccination centres use MASTA and provide authoritative advice.

Vaccinations are better avoided if you are ill or pregnant. The advice is to check with the doctor before any vaccination, if taking any medical treatment, particularly a steroid or drug for treatment of cancer, or if you have any allergies.

A vaccination programme should be planned well in advance of the trip. The time interval between different live vaccines is important. There may be reduced response if the body's immunity system has to react to more than one live vaccine at the same time; there can also be difficulty establishing which vaccine caused side effects if more than one are given together. So live vaccines are conventionally well separated (though polio and yellow fever, taking effect at different rates and given by different routes, can be given on the same day. Both are very safe. If not given together there needs to be a three week interval between them). If the second dose of a course of vaccine (for instance, for typhoid or cholera) is given earlier than recommended, the rise in antibody level and thus the degree of protection is not as great.

Cholera vaccination

Cholera vaccination by injection possibly gives a degree of protection. What is needed and as yet is unavailable is an oral vaccine to counter the cholera germ which remains in the bowel.

Cholera epidemics strike countries with poor sanitation. Cholera vaccination is valid for six months, but as stated is at best only partially effective and then for no more than a few months. The comforting play on words for the business visitor is that any one who wears a collar does not get cholera, implying such a person takes care with diet and hygiene which are of more value than the vaccine.

Vaccination does not stop transmission of cholera, but contrary to WHO recommendations, the vaccination is still required for entry to some parts of Africa and Asia and to other countries during an outbreak or at time of high risk. For maximum effect two doses, two to three weeks apart, are given. Usually one dose satisfies airport health authorities and is valid six days after first-time vaccination and immediately after any subsequent injection.

As with all diseases, spread by dirt, flies, infected food and water, cholera reflects poor public sanitation, embarrasses the government and puts off tourists. Thus there is the temptation for that country to play down the true facts: not all statistics of reported cases can be relied upon.

Cholera spreads by the faeco-oral route, through contamination of food or water with cholera bacilli, aided by flies. Those vaccinated are still at risk; they must not assume total immunity and should take the greatest care over what they eat and drink; this care gives equal if not greater protection than vaccination. The ineffectiveness of cholera vaccination led WHO to advise that cholera vaccination should not be compulsory, as it does not stop cholera getting into the country. Some countries, however, still insist on proof of vaccination, not just by travellers from infected areas, but everyone entering the country. Most of these countries accept evidence of a single dose but a few require a course of two doses. Side effects are minor, but as with other killed-germ vaccines, sensitivity, allergy to the 'foreign' protein in the vaccine, may develop after numerous doses. For maximum response, the two primary-course doses should be separated by an interval of two to three weeks. (It is said that the interval can be shortened to ten days.) Normally doses are given deeply under the skin.

However long since the first course of cholera vaccine, just one dose maximally boosts immunity. Untreated cholera is rapidly fatal in 50 per cent of cases.

Yellow fever

Yellow fever is a lethal illness, transmitted by a specific variety of mosquito in equatorial Africa and South America. Vaccination against yellow fever is long-lasting and effective, and is mandatory for visits to countries in the risk areas of Central and South America and Africa (within fifteen degrees of the Equator).

The international certificate of vaccination, dated, signed and stamped, is valid ten days after the first ever vaccination, and immediately after a booster injection given within the ten years the certificate lasts. This vaccination is only available, for a fee, at designated centres (and has to be given within one hour of being reconstituted from powder with sterile water).

Yellow fever vaccine is not given to infants under nine months, nor in pregnancy, nor (because the vaccine is grown on chicken eggs) to those few who are allergic to chicken protein, that is, eggs, meat or feathers. (The allergy is rare and many who claim to be allergic are not, as they can eat foods containing eggs such as cakes or buns.) The vaccine also has traces of neomycin and polymyxin and so should not be given to anyone allergic to these antibiotics.

There was, until mid-1978, the need with mandatory vaccinations for the doctor's signature on the international certificate to be authenticated by the local authority. The signature was compared with a specimen signature and the certificate stamped. This was to counter forgery by someone, unwilling or unable to have the vaccine, signing the certificate for themselves. Without authentication in the form of a stamp the certificate is invalid. Now vaccination certificates may be stamped at the offices of the doctor who performed the vaccination. The stamp gives the address surrounded by the words 'British Government'. A yellow fever vaccination certificate has always been authenticated at the designated centre where it is given.

Typhoid

Once typhoid vaccination was only recommended to those travelling to developing countries, but with outbreaks close by in Europe and the typhoid germ resistant to many antibiotics, vaccination is now recommended before going anywhere other than the USA, Australasia, Canada and Scandinavia. Like cholera, typhoid is a disease of poor sanitation and the same comments about prevention apply.

What was the best-known vaccine, TAB (Typhoid Paratyphoid A

and B), was withdrawn and replaced by monovalent typhoid vaccine, because there was no proven benefit from Paratyphoid A and B vaccines, which were the most likely to cause side effects.

Temporary redness and soreness at the injection site may occur, with occasionally, flu-like symptoms of fever and aching muscles. These are more likely in those over thirty-five who have had many previous typhoid vaccinations. A reaction develops within hours, lasts about two days, and if severe, is sufficient reason never to have that vaccine again. Ideally the vaccine should be injected into the non-writing arm, before a quiet weekend.

Doses are given by injection under the skin with four to six weeks between the first and second; a shorter interval reduces the effectiveness. One dose gives some protection. The chances of reaction are reduced by giving all injections, other than the first, in a much smaller dose into the skin (intradermal). It is not true that drinking alcohol makes side effects more likely. Apparently the story was put about many years ago to discourage troops drinking during their day off after typhoid vaccination, – time to allow reactions to subside.

Typhoid vaccination is effective for three years and however long it is since the previous vaccination, one dose boosts immunity to peak level.

Polio

Worldwide, there is now less poliomyelitis, but outbreaks still occur, in Europe as well as further afield, and remind us that polio is still around and not confined to countries with poor sanitation.

This illness, which can paralyse, usually affects children and is also called infantile paralysis, but it can affect adults. After a fever, arm and/or leg paralysis or worse may occur. The illness has struck unvaccinated business people abroad. Usually polio vaccination is not required to enter a country and is apt to be overlooked.

Everyone should be adequately and regularly protected against polio, particularly travellers to South America, Africa and South East Asia.

There are three types of polio virus: each of the three doses of three drops of live Sabin vaccine, given at six-week intervals, protects against one strain. Immunity is maintained by one booster every ten years. When one member of a family is vaccinated against polio, the rest of the family should be too, as occasionally the vaccine virus reverts to the wild strain and could infect another family member.

Reasons not to be vaccinated include ill health, diarrhoea, allergy to certain antibiotics, cortisone (steroid) treatment and pregnancy.

Tetanus vaccination

Everyone should be protected against tetanus, which is ubiquitous. The illness, which can prove fatal, may follow the most trivial puncture wound or nick; a lady died from tetanus after being jabbed by the fishing rod held by the ornamental gnome in her garden. Tetanus is caused by a toxin produced by the tetanus germ which damages the nervous system. Horses may excrete tetanus germs and the incidence of tetanus in the United Kingdom has fallen since the declining use of the horse for transport.

Immunity for ten years is provided by three injections of tetanus toxoid, the toxin poison made harmless but able to stimulate protective antibody formation. The second dose is six to twelve weeks after the first, and the third, six to twelve months after the second. A booster injection is given at the time of injury except in the first twelve months after vaccination. Tetanus toxoid is routinely given to babies in the triple vaccine.

Tetanus vaccination is also standard for the Armed Forces, so ex-Forces employees will have had it there too. The primary course need never be repeated. After many doses, allergy may occur. An alternative simple toxoid can then be given by injection (never to be used as a first-ever dose). It is important that dates of tetanus vaccination are recorded on the back page of the international certificate or in a personal health record which gives dates of all injections, medical examinations, eye tests, dental checks, and so on.

For protection against hepatitis A and hepatitis B see pp. 111–13. Diphtheria, Rabies, plague and typhus vaccines are available for travellers, also meningococcal meningitis A and C, Japanese B encephalitis and pneumococcal pneumonia vaccines. Measles vaccine should be given to those who have had neither the vaccine (recommended at age eighteen months) nor the illness. This vaccine is now combined with mumps and rubella vaccines.

In recent years, some lethal, highly-infectious illnesses, Marburg and Lassa fever, have occurred in West Africa. They are difficult to treat and have a high mortality rate. The background to these viral fevers is obscure. These illnesses affect the local population and medical, nursing and other hospital staff treating them. Sometimes the diagnosis all along has been malaria or typhoid fever.

Table 2 Suggested rapid vaccination schedule

Timing	Type	Notes	Dose	Certificate
DAY 1	Yellow fever	Subcutaneous (SC) injection in non-writing arm. Only available at Yellow Fever Centre. Effective and certificate valid in 10 days for 10 years. Booster effective and valid at once, if taken within ten years.	0.5 ml (SC)	Issue International Cert. of Vaccination (ICV).
	Poliomyelitis	First of course of 3, or single booster. Booster effective for 10 years.	3 drops (Oral)	
	Monovalent typhoid	First of course of 2 or 3, or single booster. First ever dose SC, 2nd and subsequent doses SC or Intradermal (ID). Booster effective for 3 years.	1st 0.5 ml SC Booster 0.5 ml SC or 0.1 ml ID.	
	Cholera	First of course of 2 or single booster. Only give if local epidemic or if mandatory for entry. Effective and certificate valid in 6 days for 6 months. Booster effective and valid at once.	1st 0.5 ml SC booster 1.0 ml SC or 0.2 ml ID.	Issue ICV
	Adsorbed tetanus toxoid	First of course of 3 or single booster. Simple version available if history of allergy. Booster effective for 10 years.	0.5 ml SC.	

	Hepatitis B	First of course of three.	0.5 ml Intra-muscular (IM)
DAY 11 to DAY 30	2nd Cholera	Only give if extraordinary risk.	1.0 ml SC or 0.2 ml ID.
DAY 42	2nd Typhoid	Can also be given during trip or after if necessary. Effective 3 years.	0.5 ml SC or 0.1 ml ID.
	2nd tetanus		0.5 ml SC.
	2nd Polio		3 drops oral.
	2nd Hepatitis B		0.5 ml IM.
DAY 84	3rd Polio		3 drops oral.
DAY 183 TO	3rd Typhoid	Only give if extraordinary risk.	0.5 ml SC or 0.1 ml ID.
DAY 365	3rd Tetanus	Effective 10 years.	0.5 ml SC.
	3rd Hepatitis B	Effective 10 years. Check effectiveness with antibody level before trip. Likely effective at least 5 years	0.5 ml IM.

Give single injection of Globulin (GG) protection against Hepatitis A one day before trip starts, at least two weeks after the last vaccination and only after checking no existing antibodies which would indicate having had the illness in the past, permanent protection and no need for GG. Vaccines also available for tuberculosis, rabies, typhus, meningitis, influenza if high risk.

The following statement of exemption from a vaccination, on headed paper, is usually accepted by the country of entry, but there can be no guarantee of this.

Medical certificate of exemption from vaccination

This is to certify that ....
has not been vaccinated against ....
for the following reason:
....

Signature of doctor and qualifications
Date
Authentication stamp

'I also certify that, to the best of my knowledge, the above-named has not been in contact with a case of ....

Followed by the doctor's signature, qualification, the date and the authentication stamp.

The health authorities of the country to be visited are not obliged to accept such a statement and can turn the traveller away or impose quarantine.

OTHER HEALTH HAZARDS

Food and drink

Immunity provided by vaccines must not lull the traveller into a false sense of security. Protection is not absolute and there are many other illnesses, such as those transmitted by food and drink, like enteritis, not covered by vaccination.

International hotels in cities, with modern kitchens, should have high standards and so should be safe but it is still best to ask, to check and to observe. Flies in the restaurant mean flies in the kitchen, and contaminated food in consequence.

To minimise the risk of gut disorders, you must take great care over what you eat and drink, especially as food and drink teeming with germs may look and taste normal. Ensure hot food is freshly and thoroughly cooked. Avoid cold meats and sausages, meat balls, curried food, 'bar tasties', salads, local ice cream, raw fish, smoked fish, shellfish and soft cheeses. Peel fruit and tomatoes. Only drink boiled milk. Don't drink or clean your teeth in suspect tap water. No ice cubes in drinks unless made from sterilised water. Even the glasses may be contaminated; if in doubt use a straw (take some with you) or drink straight from the bottle. Tap water is safer after filtering through muslin and then boiling. Water-sterilising tablets, though not one hundred per cent effective, and giving the water a distinctive taste, are simple to use and worth having for situations where boiling is not possible. A neat 'Travel well' chemical filter in a light, portable flask is available. As a last resort take really hot water from a hot tap, cover and allow to cool.

If you get diarrhoea, have lots of canned drinks and juices, adding sachets of electrolyte mixture from the 'traveller's kit', to replace the fluid loss. Eat nothing for twenty-four hours, then only bland food for forty-eight hours, such as steamed fish and certainly no milk products. 'Stopper' tablets may help if you must go out, but could prolong the condition. Diarrhoea which goes on and on, or is accompanied by a fever or is more than socially inconvenient, warrants consulting a doctor. A sulphonamide-streptomycin combination, taken throughout a two to three-week trip (provided there is no allergy) can prevent enteritis in someone who is particularly susceptible.

In essence, to prevent food poisoning, the slogan in the kitchen must be 'keep it hot, keet it cold or don't keep it'. For the customer, unaware of kitchen standards, it is best to eat only hot, fresh food, freshly cooked.

Infective Hepatitis: Type A

Infective Hepatitis is a liver disease usually caused by Type A or Type B virus. Other varieties include delta, C, and 'non A non B'. Type A, the more common in travellers, is ubiquitous and contracted from food or drink contaminated by excretion containing the virus. Swimming in sewage-polluted sea water is another way of getting this illness.

Peak incidence is in the autumn and winter. It is commonest in conditions of over-crowding and poor sanitation.

The traveller who stays at international hotels and eats hot fresh food, freshly cooked, is relatively safe. Eating out as a guest (when to say 'no' to any dish is looked on as discourteous) risks hepatitis. Shellfish are especially hazardous. About a month after being infected, the illness suddenly starts with flu-like malaise, nausea, head-

ache and aching muscles, plus discomfort over the liver. There is
rapid loss of weight and debility. The patient improves six days later
when the stool becomes pale, the urine dark and jaundice develops.
This is a deposit of bile pigment, first seen in the whites of the eyes,
with generalised itching and, curiously, an aversion to smoking.
(Jaundice also occurs in other conditions when too much bile is
produced or too little is got rid of.)

There is no specific treatment and most cases recover quite soon. A
light diet and rest are recommended and the jaundice slowly fades
over two to three weeks; in severe cases, vitamin K and a steroid are
given. Convalescence can drag on, with tiresome debility and depression. Alcohol is forbidden, as is 'the pill', for six months. Protective antibodies develop which prevent recurrence.

There is no vaccine, but one injection of human immuno-globulin
(IG) can prevent or lessen the clinical illness. IG contains the protective antibody taken from donors who have recovered. This is 'passive' immunity, that is, the recipient's immune system is not
stimulated to produce its own antibody defence. IG is at its most
effective straight after the injection, and the benefit wanes over four
to six months; so IG is given just before the trip and at least two
weeks after any vaccination to avoid reducing the effect of that
vaccine. There is most risk of hepatitis in India and the Far East, in
Afghanistan and Nepal particularly, and when off the beaten track,
as well as, of course, in any area where there is an outbreak of
hepatitis at the time.

GPs receive a fee from the DHSS for giving IG to patients travelling to India and the Far East.

Hepatitis B

Almost everyone knows how they could get AIDS – Acquired
Immune Deficiency Syndrome – but not that many realise Type B
hepatitis is passed on in the same ways, namely, by sexual intercourse, or from mother to baby before or at birth, or by transfusion
of infected blood. Hepatitis B is therefore quite different from
hepatitis A, which is transmitted in food and drink.

Hepatitis B is much more hardy than Human Immunodeficiency
virus (HIV), the cause of AIDS: it can survive drying and some
chemicals.

Hepatitis B causes illness after an incubation of two to three
months, whereas HIV can be dormant for years before becoming

active again which is when AIDS strikes. Hepatitis B can also cause sub-clinical infection, with the infected person not realising he or she has had the illness, and only getting to know of this when a blood test proves positive.

Unlike in AIDS, 90 per cent of cases of hepatitis B get better. A few have an overwhelming infection and die; up to ten per cent become life-long carriers of the virus and infectious to others. It is estimated there are three hundred million carriers worldwide. They are more susceptible to liver cirrhosis and cancer. In 1989 it is said that more people die as a result of hepatitis B each day than die from AIDS each year.

The first vaccine contained material from healthy donor carriers of hepatitis B. As the vaccine was plasma-derived, there was considerable concern about the theoretical possibility of transmission of HIV, but purification inactivates any HIV and no cases of AIDS occurred as a result of this vaccination. There is now a genetically-engineered, yeast-derived hepatitis vaccine which is much less expensive than the original vaccine and just as effective.

Three doses of either vaccine are given, the second four weeks after the first, and the third five months later. There is usually evidence of antibody protection after the second dose. It was thought pointless to offer the vaccine routinely to travellers and expatriates, but with so much 'silent' illness and so many 'silent' carriers, plus the virulence of the virus, it is now considered prudent for travellers to high-risk countries like China, South-East Asia and tropical Africa to have the vaccine. Given intra-muscularly, a better response follows injection into the upper arm muscle than into the buttock fat.

Risk of hepatitis B is another reason for ensuring use of sterile needles and syringes, and for avoiding local blood transfusion when abroad if there is any possibility the blood has not been tested for hepatitis B.

Acute hepatitis B, like hepatitis A, causes loss of appetite, nausea, fatigue and other flu-like symptoms, together with discomfort over the liver and distaste for alcohol and cigarettes. Most people with this illness notice dark urine and pale stools at first, but only a few become clinically jaundiced – in this way the cause of the illness can be missed. Once the virus has gone – usually within three months, but sometimes not until six months later – liver damage ceases and the liver is restored to normal. There is no specific treatment for hepatitis B.

Tuberculosis

Tuberculosis (TB) is passed on by close contact, and is unlikely to
infect anyone from the UK working abroad, or their families, unless
they employ local staff without first having them screened for TB.

TB is spread by inhalation of infected sputum directly or from dust.
'Primary' TB, which the body usually overcomes, often without
symptoms, protects against the more serious 'post-primary' infection.
Adults are likely to have had this primary infection, which shows up
as a small 'scar' on a chest X-ray and as a positive skin reaction to a
small dose of TB protein in the skin of the forearm (Mantoux test).

As TB is uncommon in most of the UK, children are less likely to
have had this natural protection from primary TB. For this reason
children of thirteen have the skin test and, if negative, a harmless
dose of altered TB (BCG) to develop this protection.

For younger children going to countries where TB is prevalent, the
(Mantoux) skin test should be done at the local chest clinic, and if
there is no reaction, BCG given.

Malaria

The incidence of malaria has increased considerably in recent years.
Malaria spread to such previously 'safe' countries as Algeria, Morocco and Turkey. Tragically, each year in tropical Africa alone, a
million children die from malaria.

Malaria has also returned to countries from which it had been
cleared. Not so long ago in Sri Lanka, malaria was rare, now there
are over two million cases a year.

Despite government health warnings, cases of malaria imported
into Britain have increased. In one sample year, 1452 came from
India, 452 from Africa and nine from elsewhere. 499 were immi-
grants, 158 visitors from abroad, 120 British tourists, and 105 British
business men. Nine died; all these had caught malaria in Africa. Of
these at least seven had not taken an antimalarial drug. Malaria is
spread by the bite of an infected mosquito. Some forty mosquito
species have become resistant to DDT and other insecticides.

There are four types of malaria; three develop into recurrent bouts
of fever, the fourth is rapidly fatal if not diagnosed and treated.

Thus early diagnosis and treatment are essential. Initial symptoms
are like 'flu. The doctor must ask 'Where have you been and when?'.
To misdiagnose malaria as 'flu can be a fatal mistake, as life-saving

drug treatment will be delayed. Malaria only needs one bite from an infected female anopheline mosquito: a short transit stop to refuel, a stroll in the transit lounge, is time enough for this to happen. Some travellers forget malaria, or disregard all warnings, or rely on the quinine in their gin and tonic! One small survey of sixty-nine people with malaria showed that out of the fifty-four non-European patients, only two had taken a prophylactic. Ten of the fifteen Europeans did so, but only three had intended to continue for six weeks after their return. One traveller told me, 'yes, I carry antimalarials but I only start them if I'm bitten by a mosquito!'

The antimalarial drug must be started before entering the malarial area and continued for six weeks after leaving it. One mistake is to do the opposite, to take the drug for weeks before the trip and to stop on leaving the country, assuming the danger is over. No longer is the one drug effective world-wide. In some parts, including tropical South-East Asia and Africa, malaria is resistant to several drugs. It should not be assumed that the drug taken on a previous trip will do this time – enquiry should be made about both the presence and the type of malaria in a particular area.

Paludrine, made by ICI, has no significant side effects and is available without prescription from chemists. It is still the most commonly used antimalarial agent. The adult dose nowadays has gone up from one to two tablets a day. Kept cool and dry in an airtight container, Paludrine has a shelf life of five years. When malaria is resistant to Paludrine, there are other effective drugs available.

Sometimes one drug, locally, gets an unjustified reputation of being ineffective. In a malaria area, whether an antimalarial has been taken or not, any feverish illness is rightly labelled and treated as malaria as a precaution, rather than delaying for the results of the blood test. Nothing is lost, and possible serious illness avoided, but these measures foster the myth, that the preventive drug is no use, when all the time the illness may have been a self-limiting infection like influenza.

The list of countries where there is risk of malaria for some or all of the year, in part or all of the country, fluctuates. Countries can be added and removed from the list. Check with the embassy before travelling. If there is any doubt, it is best to take an antimalarial agent.

In general, remember that mosquito bites are almost inevitable; mosquitoes favour new arrivals, especially their ankles! Use an insect-repellent lotion, cream or aerosol. Cover exposed skin at and after dusk, when mosquitoes are most active. Unless there is air conditioning and sealed windows in the bedroom, sleep under a mosquito net; soaking this first in DEET insecticide and allowing the net to dry is an added protection. DEET-impregnated ankle and wrist bands are available. These deter mosquitoes. Smoke cones and electrical devices are not really effective.

Rabies

Rabies, which features in some horror films and nightmares, is rife in Asia, South America and Africa, and occurs as close by as France. The United Kingdom has kept rabies out so far – being an island helps. During the past forty years, rabies has spread throughout Germany, Holland and Belgium, as well as into much of France, carried by infected dogs, cats, squirrels and the red fox. Rabies is passed on to humans by a bite, scratch or even a lick from an infected animal. A person can be lulled into a false sense of security by what can be a long incubation whilst the virus moves slowly along nerves from the bite to the brain. The range is from nine days after a bite on the face or neck, to over a year for a bite on the foot.

There is always time to get anti-rabies vaccine, which prevents the illness from developing, and the sooner this is given after suspected infection the better. In the past the vaccine was produced from sheeps' brains, carried the risk of severe allergy and had to be given into the stomach muscles every day – a painful procedure. Now the vaccine, available in this country from Public Health Departments, is from human tissue and carries little risk of allergy. Four injections are spread over two weeks, with boosters one month and three months later. This vaccine is also given to anyone at particular risk, like staff caring for animals in quarantine, and customs officials. The vaccine is not recommended routinely for overseas travel.

If you are bitten or scratched by an animal whilst abroad, it is important to wash the wound *immediately* using soap or detergent or just clean water. In addition, put some alcohol on the wound. If bitten by a dog or cat try to get the owner's name, address and telephone number and give them yours; ask if the animal has had rabies vaccine and if so see the Certificate. Tell the owner to contact you at once if the animal dies or is ill within the next two weeks. If the animal is a stray or wild, make a note of the animal's description and the place, time and date when you were attacked. Tell the local police. Then get medical attention fast from the nearest doctor or hospital. If in difficulty far from home, contact the nearest British Consulate or Embassy. Rabies vaccine may be given but in any event, tell your general practitioner as soon as you get back to this country.

It is almost second nature, for children especially, to stroke a pet; but when you and your family are abroad the rule must be 'Don't touch animals'.

Heat hazards

'Anything to get away from Britain's miserable, cold, wet weather' is a frequent grumble, but heat can have its snags, too. The body creates heat, and in hot weather, when there is much heat gain, keeps body temperature steady by heat loss through evaporation of sweat. The rate of loss is helped by air movement but drops when the humidity rises. Sweat is salt and water and must be replaced by fluids (at least six pints a day) and added salt on food, otherwise heat exhaustion develops. Headaches, giddiness and cramps are followed by collapse and vomiting, when hospital treatment will be needed.

In 'heat stroke', sweating stops and the temperature rises. Alcohol may precipitate this condition. The clue is that the skin feels both hot and dry. Headaches, cramps, giddiness and sickness are followed by confusion, delirium, coma and convulsions. Untreated heat stroke kills. Quickly cool the person by cold water sponging, applying ice and moving the air with a fan until the person's temperature reaches 102 degrees F. Stopping active cooling then allows the temperature to continue to fall to normal.

'Prickly heat', the commonest effect of heat, is an itchy skin condition associated with sweating, where skin surfaces rub together or skin is rubbed by clothing. Pure cotton clothing which allows air through to the skin helps, as do dusting powder and air-conditioning.

Ankle swelling is a temporary nuisance when first in a hot climate.

An aide-memoire for staff could read as follows:

To avoid problems with heat and the sun, have plenty to drink and take extra salt. Wear loose-fitting cotton clothes and avoid nylon. Apply a sun-screen generously and day by day, from fifteen minutes on day one, gradually increase the time spent in the sun (*never* the midday sun). At all other times cover up and wear a wide brimmed hat. If you do suffer mild sunburn, apply calamine cream. Severe sunburn, like any burn, is a medical emergency.

Fungal skin infections

Fungal skin infections in toe clefts, groin and ears, like athlete's foot, thrive in a hot, damp climate. Avoidance measures are to pat skin dry gently and completely after a bath, to sprinkle on anti-fungal dusting powder, and to wear clothes made of cotton, and leather-soled shoes.

AIDS and overseas travel

AIDS (Acquired Immune Deficiency Syndrome), caused by the Human Immuno Deficiency Virus (HIV), is a possible hazard for travellers who are injured or taken seriously ill in countries where standards of medical care are suspect and emergency treatment is needed by way of injections of drugs, suturing a wound, or the giving of intravenous fluid. Injections may be given with an unsterile needle or syringe contaminated with HIV-positive blood; likewise the needle used to sew up a wound or the needle used to give intravenous fluid or, last but by no means least, blood transfused may contain HIV. Because of this risk, travellers can take sterile equipment with them in a first aid kit, to hand to the doctor treating them, plus a bulky blood substitute which can be used in an emergency whilst arrangements are made for immediate repatriation by air, or for HIV-free blood to be flown out – arrangements have been made for the National Blood Transfusion Service to supply this for a UK citizen overseas. Alternatives are to have your own blood stored for such an eventuality, or to use a local safe donor panel of fellow expatriates. It is of course essential to know your own blood group, either by becoming a blood donor or by a specially arranged test. The snags of an AIDS kit include suspicion by customs officials of drug abuse, of not having the kit to hand if the need arises, or of not being in a fit state to suggest its use.

In countries where there is a high incidence of AIDS and therefore of people who are HIV-positive (for every case of AIDS, there are a hundred people who are HIV-positive but appear quite healthy), the risk of contracting the disease from a sexual relationship with a local person is very great. This is far and away the greatest risk to a traveller. It makes sense to avoid an injection. This is one reason for the traveller to be dentally fit. There is no risk from an insect bite or from giving blood, nor from food and drink, from cups or glasses, from lavatory seats or swimming pools.

Some countries require either evidence of a recent negative HIV-antibody test before granting an entry visa, or insist on such a test immediately after entry to the country. In either situation it is essential for there to be skilled counselling before and after the test is performed. It usually takes three months from infection for the test to become positive.

Whilst on the subject it is pertinent to mention that some companies include a question about having AIDS or HIV in the pre-employment medical questionnaire, and a few companies carry out a screening antibody blood test before appointing a person to the staff. Employment of a person with AIDS, or who is HIV-positive, will not endanger the health of other staff, but some employers want to avoid recruiting and investing heavily in the training of a person who is likely to become ill and subsequently die. Statistics have shown that half of all AIDS patients die within a year of diagnosis – drug-abusers sooner than homosexuals. However, seventy-two per cent of the latter group, who presented with Kaposi's Sarcoma rather than an infection, lived more than a year, and thirty per cent were still alive at least five years from diagnosis. As yet unknown is the true percentage of HIV-positive people who will go on to develop AIDS. It could be that this will be one hundred per cent.

TRAVELLER'S MEDICAL KIT

Travellers should carry a medical kit. This could be provided by the company as part of the health care scheme. Lightweight and small for easy packing, the kit is not comprehensive but has treatment for common conditions as well as the usual plasters. Of course, for any illness which seems at all serious, a fever for example, the message is 'don't delay by trying a first aid remedy, get to a doctor at once.' (In an attempt to overcome language difficulties, the Health Education Authority has produced an illustrated multilingual Medical Phrase Book in English, French, German, Spanish, Italian and Hindi that covers 150 medical and surgical situations.) In any major city, the company should know of an English-speaking, reliable doctor to be contacted.

Kits which contain medicines must not be given out by the personnel department to whoever is travelling, only a kit which contains dressings only. Each kit with medicines is prescribed by a doctor for an individual, and a check will be made to exclude allergy or interaction between any drug in the kit and an existing condition or treatment.

A dressings-only kit contains:

1 3 inch crepe bandage and safety pins
6 medi-wipes
6 sterile dressings; a mixture of small, medium and large
6 sterile waterproof plaster dressings
1 pair of disposable scissors
1 clinical thermometer
1 short first aid 'aide-memore' on a card.

(It is also useful for any traveller to know the rudiments of first aid. The company nurse can teach these in a single two hour session). Such a kit in a plastic box is available at any Chemist. To create a traveller's kit tailor-made for the individual and the countries he or she is to visit, some of the items may need a doctor's prescription and may include travel sickness tablets, tablets against diarrhoea, tablets for pain relief, antacid for indigestion, cream for bites and stings on unbroken skin, insect repellent, short-acting sleeping tablets and water sterilisation tablets.

Personal instructions for use of traveller's kit (to be included in the kit)

This kit contains treatment for minor ailments overseas.

The drugs are intended for essentially healthy adults, and are not to be taken by children or in pregnancy.

If you are on drug treatment, carry with you details of the drug, the dose and a covering letter from your own doctor. Check with your doctor that it is safe for you to take the drugs in this kit as well as your own.

Tablets can deteriorate and have a limited shelf life. You must return this kit, used or unused, to the occupational health centre at least once a year for restocking and replacement. This is important.

Pain – paracets tablets (6) These contain paracetamol. Take two tablets initially and, if necessary, repeat the dose after six hours. Paracets are suitable, for example, for a headache, toothache, a pulled muscle. If in severe pain consult a doctor.

Sleeplessness – normison capsules (4) Take one capsule on going to bed. Don't drink alcohol or drive for twenty-four hours.

Diarrhoea – imodium capsules (12) – diorylate sachets (6) Take two capsules and one sachet dissolved in bottled water, after every loose stool – no more than eight capsules in twenty-four hours. Also drink plenty of soup, tea, bottled water or soft drinks. If diarrhoea is accompanied by a fever or repeated vomiting, or persists acutely for more than twenty-four hours, consult a doctor.

For minor wounds apply direct pressure to stop any bleeding; clean with *mediwipe* (4). Cover with an *adhesive plaster dressing (6)*. Use *melolin dressing (2)* and *bandage (1)* for a larger wound. If the wound gapes, consult a doctor.

Upper Respiratory Tract Infection If possible avoid flying; if you have to fly, use a *decongestant nasal spray* – one spray in each nostril just before take-off and again before landing.

Mild Indigestion – Gelusil Tablets (6) Suck one tablet as required.

Insect Bites/Stings Avoid these by using '*shoo wipes*' (3). If you are bitten, apply a very small quantity of the *hydrocortisone cream*. Do not scratch.

For a visit to a malaria risk area, antimalarials are added to the kit. A person on a short trip who is susceptible to diarrhoea may be given an antibiotic mixture which may prevent an attack. In a cold climate a lip salve is useful. Several of the items are not strictly first aid but can make life more comfortable, as can a pair of earplugs to use if the hotel is noisy at night, a door stop for security, and a bath plug. Tablets go off and deteriorate faster in wet heat so, as stated, the kit must be checked regularly by the occupational health department, and deteriorated and out-of-date items replaced.

FLYING

Many people don't like flying – the take-off is especially stressful – but are able to handle their fears. A few are terrified and will not fly: just the thought of it causes a stress reaction of fast pulse, sweating, shaking and sickness. They fear the possibility of a crash, of being taken seriously ill, or being trapped in an enclosed space.

Reluctant, ashamed or embarrassed to admit a weakness, such people may pass up promotion opportunities which involve flying. Yet behaviour therapy from a psychologist can be quickly effective- it consists of the teaching of relaxation in a progressively more personally stressful situation, simulated by recordings and even a mock-up plane, before an actual flight with the psychologist.

People should not fly if they:

(1) Are suffering from a grave, that is terminal illness.
(2) Have an infectious disease.
(3) Are more than thirty-five weeks pregnant.
(4) Are suffering from anaemia (because of the reduced oxygen pressure).
(5) Have a chronic lung disorder (because of the reduced oxygen pressure).
(6) Are in heart failure.
(7) Had a heart attack less than a month ago.
(8) Had bowel surgery less than ten days ago.
(9) Had chest surgery less than three weeks ago.
(10) Had a bleeding ulcer less than three weeks ago.
(11) Have had a recent stroke.
(12) Are suffering from acute earache, sinusitis, or other upper-respiratory-tract infection.
(13) Have an acute mental illness (unless they are well-sedated and adequately accompanied).
(14) Have poorly-stabilised diabetes.
(15) Suffer from poorly-stabilised epilepsy. Treatment should anyway be increased for the flight.

If in any doubt about fitness to fly, check with your doctor and with the airline concerned. The airline must be told in any case, and relevant forms completed. Airport formalities can be speeded up, a wheelchair or an ambulance provided to meet the flight. The pilot has the final say about who will fly.

TIPS FOR TRAVELLERS

To counter travel sickness a number of effective preventives in tablet form are available, some on prescription only, to be taken one to two hours before the journey. Drowsiness is a common side effect, intensified by alcohol which, along with driving, should be avoided for that day. 'Sea Bands' in which a button presses on the acupressure points below the wrists can also be an effective preventive and has no side effects.

It is best to lie down (a row of empty seats on the plane may allow this) and ideally go to sleep. Sitting for hours on end impedes circulation in the legs. There is swelling and the possibility of a clot in a vein, so wear loose-fitting clothes and comfortable shoes; loosen shoe laces; do not sit with crossed legs; get up and walk about from time to time. A prebooked aisle seat better allows for this.

The cabin is pressurised to the equivalent of between 5000 and 8000 feet. There is marginally less oxygen at this height. This does not matter to the healthy, but to anyone with a lung or heart disease, it can make all the difference. Smoking makes the situation worse, as the carbon monoxide in cigarette smoke is taken up by the blood rather than oxygen, and smoke-filled air irritates the lungs.

It has been claimed that 'ozone sickness', occurs in airline passengers. Ozone, produced by the effect of ultra-violet on oxygen, is in the upper atmosphere. It is largely destroyed by the plane's air-conditioning system. Theoretically any ozone problem would be greater on supersonic flights (20 000 feet higher than other jets), but in practice there has not been a problem.

Exposure to ozone in concentrations of three ppm (parts per million) for over three hours leads to throat irritation, coughs and shortness of breath. These symptoms have been reported by passengers and crews of American airlines, but British experts consider the cause to be a combination of a long flight and dehydration caused by inadequate humidification of cabin air.

It is better not to fly with a cold or other respiratory-tract infection, because congestion blocks the Eustachian tubes which link the ears and throat, and prevents equalisation of pressure between the middle ear and the atmosphere on ascent and descent, resulting in pain in the ears, temporary deafness, and possibly aggravating any infection. If you have to fly, decongestant nose drops or a spray may keep these congested tubes patent; chewing and swallowing help too.

The reduced pressure makes body gases expand. This is especially noticeable in the bowel. Fizzy drinks give a 'blown-out' feeling and are best avoided; for the same reason only eat a light meal.

Do not fly within ten days of a bowel operation, because the expansion of gut gases strains healing wounds. After a lung operation, three weeks should elapse before flying.

Air travel can be stressful for the healthy as well as the sick traveller when many time zones are crossed; this temporarily upsets the body's 'circadian' rhythm, leading to 'jet lag'. There is a twenty-four-hour cycle of, for example, hormone output and body temperature which ensures efficiency by day followed by recovery during sleep by night. Body temperature peaks in mid-afternoon and falls to its lowest at about 2 a.m. Bowel activity stops at night. Hormone production also follows a pattern. Cortisol secretion peaks at 8 a.m. and is lowest at midnight. Shift workers may be similarly affected when on night shift, with insufficient time to adapt before they switch back to a day shift. The body's system takes several days to adjust to a new timetable, meanwhile acting as if it were still the night, when in the new country it is daytime.

Travelling in a mainly North-South direction with little or no time change does not upset 'circadian rhythm', but the unfamiliar new surroundings and the journey itself, added to sleep loss on a night flight, will be tiring. East-West and particularly West-East travel with over five hours time-change causes jet lag which can take up to five days to clear. Symptoms include apathy, tiredness, irritability, impaired concentration and disturbed judgement. This is at its worst in the first twenty-four hours. There is considerable variation and it seems that the inexperienced traveller is more susceptible.

Experienced travellers have probably learnt by trial and error how to minimise the effects. Someone new to the job, excited by the glamour and challenge, and who wants to 'get on with it' on arrival, is more susceptible. This is especially likely if the journey is West to East, shortening the day. Enthusiasm may help to sustain people, but there is the risk of failure, especially for the middle-aged. Apart from a busy round of visits and meetings in a short time, well-meaning hosts may lay on receptions and parties from the first night, which can be more tiring than work.

So, when crossing many (five or more) time zones, prepare for the trip in good time, have two good nights' sleep before the flight; start to adapt to the new times during the flight (for instance, change your watch to local destination time as the plane takes off); arrive near to the normal bed-time of the country being visited; check into a hotel, and go to bed. If possible, have the next day off, free from business meetings and lavish hospitality. Return to the UK just before the weekend – to be spent at home. Don't cram in as much work as possible from the moment of arrival, however short and expensive the trip.

One way round jet lag from a trip from the United Kingdom to the east coast of America is to fly Concorde there and back on the same day.

Sufficient sleep is vital; lack of sleep upsets all the planning in the world. Most people need between seven and eight hours sleep each night, some six hours or less, and a few nine hours or more. Children and teenagers need more sleep than adults and the elderly rather less. Loss of more than two to three hours' sleep for more than one or two nights impairs efficiency and concentration. Having no sleep disrupts thought processes. The traveller should try to sleep without sleeping tablets, but it is a sensible precaution to have a small supply just in case sleep is difficult for the first night or two, when the body will be acting as if it were day-time, with all systems tuned for peak perform-ance. Doing this will avoid waking, as I did, at 2.30 a.m. during the first night in New York.

Alcohol increases the effects and the after-effects of sleeping tablets and so should be avoided. Any next-day drowsiness precludes driving a car, particularly driving on the other side of the road. Never take a sleeping tablet on a plane as there is the slight risk of transient global amnesia, remembering no more until in your hotel room many hours later.

Digestion is also upset by travel. Light, easily-digested foods should be eaten both on the plane and on arrival and it helps to miss an occasional meal during the first day or two.

An initial company-paid telephone call home should be encour-aged, as the security of knowing all is well at home, and vice versa, is important.

Tour-end repatriation stress can be as bad or worse than expatriate stress on arrival overseas. A medical examination can be helpful and includes where relevant a screen to exclude any tropical infections.

Staff should be advised 'If you become ill after your return to the UK and need to consult a doctor, be sure to tell the doctor where you have been and when'.

INSURANCE

No one, however healthy, should travel far without being insured against illness and accident; medical treatment abroad can be expensive. (In this country, cushioned by the NHS, the true cost of medical treatment can be under-estimated.)

The NHS does not provide for illness abroad and so no drugs, other than small quantities of what would be required anyway, may be prescribed. Simple remedies for pain and indigestion can be bought without prescription.

Depending on the country to be visited, this insurance cover will be either for all medical bills or for any additional cost where there is reciprocal arrangement with the United Kingdom or partial payment by an EEC partner.

Some Private Patient insurance schemes also cover treatment abroad and offer additional optional insurance to fill any gaps.

In EEC countries, free or subsidised emergency treatment is available under the country's health service. Procedures differ in a bewildering way between countries.

In Eire all that is needed is one's signature to a statement of eligibility; in Denmark there is also the need to show one's passport. In Belgium, France and Luxembourg, payment usually still has to be made at the time for part or all of the treatment, but can later be reclaimed in full or in part from the DHSS Overseas Group, Newcastle upon Tyne, NE98 1YZ. Always get receipts for medical bills.

Leaflet SA28, 'Medical Treatment for Visitors to EEC countries', from local DHSS Offices and travel agents, gives details and includes form CM1 to complete and send off to obtain Form E 111. This must be taken on the trip and is valid for six months. Free treatment for accidents, as well as for illnesses which start after arrival, is also available on production or either passport or NHS medical card or both in the following countries that have reciprocal arrangements; Austria, Bulgaria, the Channel Islands, Czechoslovakia, Gibraltar, Malta, New Zealand, Norway, Poland, Romania, Sweden, the USSR, Yugoslavia and Australia. In Norway and Sweden 75 per cent of fees paid for medical treatment can be reclaimed.

REPATRIATING THE VERY SICK

If repatriation of a member of staff or their family is necessary, as for instance when the appropriate specialist or treatment is not available locally, or the climate aggravates the illness, and the invalid is to travel by air, clearance must first be obtained from the airline, which needs all relevant details.

The airline will make whatever special arrangements are necessary. For example, oxygen can be provided or seats removed to accommodate a stretcher. Aircraft carry first aid kits and stewards and stewardesses have first aid knowledge.

Occasionally an airline will refuse to fly a sick person and in all cases the Captain has the final say about who he will or will not take. An air ambulance is an alternative but can be very costly.

Those seriously ill must be accompanied by a doctor and/or nurse.

Airport Authorities may allow special arrangements for boarding from an ambulance, disembarkation into an ambulance on arrival, and early completion of customs inspections and emigration and immigration formalities.

9 Retirement and Redundancy

RETIREMENT

The UK population had been around 54 million for some years, but within that total, the prevalence of the different age groups is changing, with many more older people. In 1985 one in seven was over sixty-five. More than half a million people retire each year. Retirement age from full-time employment is not standard and spans a wide range. In the UK, state pension starts at age sixty-five for men and age sixty for women, although statistically women live longer than men.

With the emphasis on equal opportunities and sex equality, there is pressure for retirement age to be the same for both sexes.

Companies either retire staff at the same age as when the state pension begins, or earlier. Sometimes retirement is after a certain number of years' service. It is becoming more common for men to retire at sixty, but just as some companies in recent years have brought down retirement age from sixty-five to sixty, others have gone the opposite way raising retirement age to sixty-five. Retirement age for women has been as early as fifty. Similarly, such early retirement is standard for certain ranks in the armed forces, and for the police, psychiatrists and staff who have worked in the tropics. On occasions staff may be able to retire early, but by doing so they lose a percentage of their pension for each year early they retire; this is often referred to as 'a penalty', but is more likely a modest actuarial deduction.

Most people, if asked in early middle age, say they are looking forward to retirement – 'can't wait' – but as the event approaches, some are less enthusiastic, though usually putting on a brave face, and a few become depressed, whilst others become 'demob happy' and neglect their work in the closing weeks. People who cannot face up to the inevitable, and close their minds to what will happen, put off any planning for retirement; they leave it to the last minute, or neglect the subject altogether. This way spells trouble to those to whom the job is everything and who will especially feel the loss of position, prestige, power and the sense of belonging, which retirement removes upon them. These people in particular need help by way of positive, repeated encouragement, urging them to have a programme for planning for retirement. Reactive depression subsequent to retirement is quite likely in a vulnerable, unprepared person who will become shut in, avoid previous social contacts and activities, and need medical help. There is little scientific data about such reactions and surprisingly little known on a national level about what happens to those who have retired, but it would appear that those who deliberately choose to retire early, when questioned later, are delighted to have made that decision. Usually they have carefully planned how to spend their time. Also their plans have included an accurate financial budget, so money worries, the major concern for many retired people, are less likely.

The first month after an unplanned retirement can be very disturbing. The claim has been made, but not substantiated, that the sudden 'cut-off' trauma of retirement can be lethal and that a significant proportion of retired people die within three years. This does happen to some widowers in three years after bereavement, but whilst retirement can be likened to a bereavement, there is no hard evidence that it carries the same degree of mortality. There are, of course, anecdotes of individuals dropping dead within days or weeks of retirement, but these are totally unrepresentative. Retirement may last a long time – a third of a lifespan or more – and taking retirement age at sixty-five, on average a man will live a further twelve and a half years and a women a further sixteen years. In my experience, the majority take up other work, paid or voluntary, full-time or part-time, within months of retirement.

Sometimes a company has little or no further contact until notification of death, and it is significant that many of those least satisfied with retirement are from companies which do not maintain contact.

Other companies have schemes for keeping in touch with former employees through Pensioners' Associations' newsletters, meetings and reunions. A company with a national network of offices or stores can arrange for each unit to maintain contact with pensioners in that area and be available to help in time of need.

A few companies have links with residential homes for the older retired person or spouse, and a handful of large companies have built their own residential homes run by trained staff. Although it is realised that disability may occur in the future, criteria for acceptance may include being reasonably able-bodied and independent rather than needing 'long-term nursing care'.

To some people, the concept of fixed-age, total retirement is artificial, with its sudden switch from total work to total leisure. They advocate a gradual decrease in the time spent at work, spread over several years. The five-day week becomes a four-day week, then a year later, a three-day week, and so on; the other days are spent developing skills to use in retirement.

Such a scheme is only practicable for staff involved in routine work. It is also expensive from half-way since two people are then needed for one job.

The aim should be to develop interests in many areas well before the day of retirement. Then, should accident or illness limit activity in one field, there are alternatives. Another idea is for retirement age to be flexible; and another novel idea is for retirement for five years at fifty with a return to work until seventy. There are, of course, the practical difficulties of filling the position meantime, and of the employee getting 'out of touch'.

Reluctance to face the prospect of retirement makes it that much more important for the company positively to encourage planning for retirement. Guidance needs to be given sufficiently early to be put to the best use. Some say it is never too early to start, but it could be off-putting to anyone in their fifties, who considered themselves to be approaching the peak of a career, to be told to start planning for retirement, unless this was put in a most tactful and discreet way.

On the other hand, to give out a book on planning for retirement, along with pension details and a farewell letter of appreciation, a few weeks before the date of retirement, gives little time to put the advice into effect. A compromise is to start the programme two years before the proposed retirement date.

A few companies with the necessary expertise hold their own voluntary pre-retirement courses in company time; others arrange for an outside organisation such as the Pre-Retirement Association, involved with all aspects of pre-retirement education, to run the course. Several organisations offer a two-day course. A course may be concentrated into these two days or spread over weekly or monthly sessions, each dealing with one topic. Other companies rely on providing a book or loose-leaf binder filled with information on planning for retirement. A snag with any book is that not everyone can get information from the printed page, there is often a need for direct personal contact.

The spouse, who is of course directly involved, is also usually invited to attend a course. The housewife is unused to having her husband, possibly unhappy, frustrated and bored, at home all day, every day. I like the reported reply from one wife who, when asked for her view on her husband's retirement, said 'I married him for better or for worse, but not for lunch every day'. A working wife may not like the prospect of her husband becoming more dependent on her earnings.

Structured discussions including a question and answer session are more likely to establish rapport than formal lectures, reminiscent of the schoolroom. The talk of health, usually but not necessarily given by a doctor, is always popular. The speaker should know beforehand the ages and status of the group and, by means of written questions, what in particular they want to hear about. To be avoided at all costs is a depressing monologue of illnesses, one of which is bound to 'carry away' each member of the audience sooner or later. The informal talk should look at likely problem areas but also emphasise how to preserve good health.

A short, general talk on life style is followed by specific answers to the special topics raised and then a discussion; in this way the audience is likely to participate. Understandably, some of the questions will be about personal health problems. Other subjects/speakers for a pre-retirement course include:

A social worker to outline social service facilities and available benefits.

A dietician to discuss requirements for a balanced diet and the relationship between diet and health.

A representative from the DHSS to explain qualifications for state benefits.

An accountant to discuss savings, investments, pensions and so on.

A talk about further work opportunities, paid or voluntary.

'Geriatric' is a term to be avoided and is regarded by most retired people as an insult. Old-age pensioner (OAP) also grates; few retired people of any age regard themselves as old. 'Late middle age' was the answer from one eighty-four year old when asked at what stage of life he regarded himself; another of similar age, conceding he was elderly, added 'but young at heart'. Senior Citizen is a more acceptable title if one need be used at all.

Half the people aged between sixty-five and seventy-four have no disabling health problem, eight out of ten in this age group consider their health good, and of those over seventy-five, a third have no significant disability and three out of five consider their health good.

It is interesting to compare these findings on population health with those of the survey amongst adults under sixty-five, commented on in the chapter on sickness absence. One sad fact is that dementia affects twenty per cent of over-eighty year-olds, creating a burden for the carers. Depression is also common. The GP is paid more to look after an older person, so there need be no hesitation about regularly consulting a GP – 'it's your age' is an unsatisfactory explanation for a symptom. When given this explanation for pain in one knee joint, a seventy-five year-old retorted, 'then why doesn't the other knee hurt the same – it's just as old'.

A pre-retirement health check is a much appreciated gesture a company can offer shortly before the retirement date, providing reassurance as well as positive guidance before the step into the 'third age' of retirement. The check could be carried out by the company nurse. It includes blood pressure, a urine test, eyesight and hearing checks and weight measurement, followed by the opportunity to discuss the results and any particular health problem, with results sent to the GP.

To move house and location at the time of retirement, uprooting from the familiar surroundings of many years and leaving friends and relatives, adds another stress to that of stopping work. A country cottage, honeysuckled and romantic in summer, may be bleak and lonely in winter. Seaside resorts often attract old people, but these communities can become unbalanced with a preponderance of the elderly. A move to a smaller house when the children have grown up and left home, with less to clean and maintain, or to a bungalow if stairs are difficult, is, however a sensible decision.

To put difficulties over retirement into perspective, I should point out that most people drawing a pension whom I see are well into retirement, and very happy and content to be retired, retaining a keen interest in the company for which they worked. Some have the healthy ambition to live long enough to have taken more years of pension out of the pension fund than they contributed to it!

REDUNDANCY

If retirement at the end of a career can be traumatic, how much more likely is this reaction to redundancy. Quite apart from the money, people need to work. There is often little warning of redundancy and the suddenness of the shock adds to the stress. The middle-aged are particularly hard hit. The young person has less at stake and, from an age point of view, a better chance of finding other work; the older worker may be offered early retirement; but the average forty to fifty-five year-old has much to lose from having to leave the company and is likely to have difficulty getting another job. Added to these difficulties and the financial worries, is the demoralising loss of face with friends and neighbours from being made redundant and put out of work. Typically, after the initial dismay, there is a period of optimism that turns, if another job is not soon forthcoming, to pessimism and gloom, apathy and a fatalistic attitude. The feeling of being superfluous, unwanted and rejected and of being stigmatised, explains why occasionally a husband who has been made redundant, says nothing about this to his family, and continues to leave the house each morning as usual. The stresses of redundancy affect the most stable person, and a vulnerable personality could break down, especially if the person is middle-aged and having to face the fact that ambitions will not be realised. Providing an opportunity for counselling by a clinical psychologist, who will also teach relaxation and coping skills, helps ease the stress of redundancy and can be incorporated into the package provided by the employer, along with advice on getting another job, obtaining redundancy pay and so on. The employer may use independent outplacement consultants who provide a detailed programme. This first task is to help the former employee to come to terms with what has happened, then to work out strengths and weaknesses and decide what type of work to go for. Next the C.V. is prepared and interview technique polished. A base is provided during the job search. Successful placement usually takes four months. If the job doesn't work out within a year they start again at no extra cost.

EARLY RETIREMENT ON HEALTH GROUNDS

Company policies on early retirement vary. Those with the need to 'slim' the work force may encourage early retirement to speed up natural wastage and avoid redundancies. Others, concerned to protect the pension fund, only allow early retirement in exceptional circumstances and impose an actuarial percentage reduction of pension entitlement for each year remaining to normal retirement. Having to stop work on the grounds of ill health can happen at any age, but as would be expected, the numbers of employees leaving in this way go up with increasing age. Early retirement can also be an attractive alternative to management for dealing with an unsatisfactory employment situation and the one least likely to cause aggravation.

Numbers taking early retirement on health grounds increased significantly in 1983, probably associated with job release schemes for older workers.

Some doctors recommending individual patients to retire early on the grounds of ill health, write to the company saying so; but the terms for stopping work are, strictly speaking, not for them to say; they can only say they think their patients should stop work. The granting of early retirement is a management decision. What a doctor is being asked for is an opinion about whether their patient is medically fit or unfit (a) to continue work at all or (b) to continue a particular job or (c) if medically unfit for this, what type of alternative (for instance, sedentary work), he or she is fit to do, should such work be available. In this latter situation, when the employee is medically unfit to continue his or her particular job, management will look into the availability and feasibility of any alternative work and if this is unavailable or inadvisable, will decide on the terms for that person leaving. The GP or Specialist may well have little or no idea of what other work is available or what the patient's usual job involves in any case, and this is where a company doctor, fully versed in the activities of the company, can be most helpful.

Early retirement is voluntary and needs the active agreement of the employee and an application in writing to avoid any later risk of a claim for constructive dismissal. Without this, early retirement is impossible. The alternatives are termination of contract and resignation. A few doctors seem to think that retirement and resignation mean much the same, whereas with resignation a pension is deferred until normal retirement age; on early retirement of ill health or any other reason, the pension starts immediately. Where retirement on the grounds of ill health is indicated, but the individual will not hear of it, then the company has to terminate the contract of employment. All possible alternatives must have been considered and the member of staff be kept fully informed and given the opportunity to discuss the matter.

In some companies, the letter sent to the member of staff agreeing to early retirement on the grounds of ill health includes a statement to the effect that if health improves sufficiently, the company may require that person to return to work, in other words, the company retains this option. My understanding is that in practice this is rarely if ever taken up.

A person who retires early on health grounds may well have difficulty finding another job should his or her health improve sufficiently to allow this. Employers are understandably cautious about taking on someone who has retired early from another job on health grounds. For this reason, if in the medium term there is the strong possibility of curative treatment and the person then wanting to work again, it is better for that person to resign rather than retire. This would also apply where the condition prevents a long commuting journey on public transport, but would allow a job local to home. Again, where a particular complaint such as recurrent backache makes a manual job impossible, a sedentary job may be a practical possibility. In order to justify early retirement on health grounds, the condition must be likely to last indefinitely with no likelihood of marked improvement or cure. To retire someone early on lesser grounds can lead to the anomaly of the 'young' pensioner recovering to work full-time again elsewhere, to say nothing of the drain on the pension fund.

Medical indications for early retirement may not be clear cut. In some instances, for example severe arthritis of a knee or hip, a city clerk will be able to work when actually there sitting at a desk, but as commuting by public transport in the rush hour becomes an increasingly painful ordeal, it will become necessary to give up work. A large company with many branches may be able to arrange transfer to a local office, allowing commuting by car and thus employment to continue. Otherwise if he or she were to work locally for another company, this would be another example of the seeming anomaly of a person being retired early from one job on the grounds of ill health and yet soon afterwards taking another job and gaining considerable financial advantage: a pension plus a new salary. The success of a new treatment would be another solution. In the example given, this would mean replacing the diseased joint with an artificial joint. To someone with a grave condition, such as disseminated cancer, which carries a poor prognosis despite treatment, the prospect of getting back to work may be the goal that keeps him or her going. In almost any illness, however poor the prognosis, a doctor will leave some hope, and this may give the patient a false impression that a return to work will be possible. The company which, is made aware of the true prognosis by the company doctor, may be able to go along with this. Such cases need especial sensitivity and tact. If the goal of the dying patient is to return to work, mention of early retirement can destroy morale. Finding out 'who knows what' is an essential prerequisite. Close liaison, co-ordinated by the company doctor, with the relatives and doctors, is necessary to ensure the best financial deal for dependants without affecting morale. For example, continued sick leave and death in service may have financial advantages over early retirement and immediate pension based on years of service. Concern for the individual's needs must be balanced against the need for continuing company efficiency, and of course paid sick leave cannot continue indefinitely.

An accurate prognosis can be impossible for a particular case. Though management would wish otherwise, predicting future health is, at best, imprecise, a balance of probabilities. Predictions in psychiatry are particularly uncertain. The prognosis given by a psychiatrist is often rightly optimistic, but in some cases I have found the risks of recurrence understated, or the time to full recovery from a particular bout of nervous illness underestimated. In acute psychiatric illness, long-term decisions about future employment should, if possible and practicable, be delayed until recovery. The exception is if work alone caused the illness, in which case, a change of job location or type of job within the company may allow a return to work. People with depression have low self-esteem and may offer to resign or be demoted. Work should never be allowed just as occupational therapy. The company must remain viable and profitable and any job must be done properly.

Early retirement for a young or middle-aged person because of psychiatric illness hinders future job prospects, for when and if there is sufficient recovery to allow work of some kind, prospective employers may be reluctant to take that person on.

THE JOB RELEASE SCHEME

The Job Release Scheme was a Government measure to reduce unemployment. The first scheme covered men of sixty-four and women of fifty-nine in full-time employment. The scheme was then extended to also include men aged sixty-two to sixty-four and disabled men aged sixty to sixty-four, defined as those who on account of injury, disease or congenital deformity are substantially handicapped in obtaining or keeping employment suited to their age, qualifications and experience.

These people had the opportunity to apply to retire before statutory pensionable age, provided their employers agreed. The employer had to recruit replacements as soon as possible from the unemployed register, but not necessarily to the same jobs.

A disabled man, if not registered disabled, submitted with the application form a medical report form completed by the GP.

If it was a disabled worker who had retired under the scheme, the replacement had wherever possible, to be a disabled person.

Part II

The Office Environment

10 Working Conditions

THE OFFICES, SHOPS AND RAILWAY PREMISES ACT 1963

Despite HASWA, much of the contents of the Offices, Shops and Railway Premises Act (OSRP) continue to apply, covering many aspects of the working environment including work space, cleanliness, seating arrangements, lighting, passageways, floors, stairs, and provision of toilets, washing facilities, drinking water and personal lockers. The OSRP Act covers some of these topics in detail and others in outline.

The employer must still either display prominently an abstract of the OSRP Act and Regulations (poster OSR9, 30 x 20 in, from HMSO), or give each employee an explanatory booklet, OSR9A or 9B, within four weeks of starting work.

HASWA also expects the employee to report to management without delay equipment which is in any way faulty: Section seven of this Act states that employees will take reasonable care of themselves and others in the course of their employment, and co-operate with their employer in enabling the employer to carry out his or her other duties under HASWA. In other words, whilst the employer is required to provide safe equipment, safe systems of work and safe premises, as well as arrange adequate training and supervision (in the implementation of safety policies), employees must co-operate to meet the provisions of HASWA and take all reasonable care to avoid endangering or injuring themselves or others whilst at work.

WORKING TEMPERATURES

Section six of the OSRP requires the employer to ensure a reasonable temperature at work, and details minimum temperature requirements for work areas where people are employed for other than short periods, and where most of the work does not involve severe physical effort. After the first hour, a reasonable temperature is not less than a still chilly 16° C/60. 8°F; the ideal is around 20°C. A thermometer must be provided on each floor so employees can check the temperature. Rooms used by the public where maintenance of this temperature is not reasonably practicable or where it would cause goods to

139

deteriorate, are excluded. In these areas, employees must have access to means of warming themselves, and regular opportunity to do so.

There is an unwritten rule that some leniency is allowed on Monday mornings when the building has been empty over the weekend and that it may take longer than an hour in winter to warm the offices above the required minimum temperature.

Difficulties can occur when air conditioning has been designed for open plan and this layout is then split into individual, enclosed offices. In some of these, the air conditioning may be too strong and in others too weak, always bearing in mind that air conditioning tends to be an 'Aunt Sally'. A difficult time are the first days and weeks in the life of a new building whilst the air-conditioning system is balanced, location of thermostats sorted out, and so on. It is not unknown for a thermostat to be wrongly connected – or not connected at all – or to be behind a coat-stand.

Freestanding supplementary heaters can be an accident risk and so are discouraged, but wall-mounted air-cooling units, operated by the staff when necessary, can be very useful where the air conditioning was not designed for the amount of electrical and electronic equipment that has since been installed and which creates extra heat.

No maximum temperature is given, by the Act, and on a hot, sticky summer afternoon, in a non-air-conditioned office facing south with lots of glass, working conditions may get very difficult. Opening windows doesn't help much – dust, dirt and traffic fumes blow in and traffic noise is a distraction. Providing fans shows that something is being done, but the benefit is partly psychological, giving an impression of coolness whilst circulating existing warm air; whilst heat is lost from the body by evaporation of sweat, the temperature of the room may actually rise. A special plastic film for windows reflects heat and cuts glare and temperature.

SICK BUILDING SYNDROME

In the so-called 'Sick Building' Syndrome (SBS), a high proportion of staff in what is often an air-conditioned (a/c) building suffer relatively minor symptoms – minor in the sense that they might not cause sickness absence, but distressing and detrimental to productivity. These symptoms include headaches, sore throats, and sore eyes. Measurement of, for example, room temperature and humidity are necessary; classically in SBS all tests are normal. Regular hygiene maintenance of cooling towers, if present, will exclude any question of Legionnaire's Disease. In such a building, I find there is likely to be an increase in the level of carbon dioxide in the air. This is the gas we all breathe out, which is barely present in the fresh air we breathe in. A rise suggests either insufficient fresh air getting to the rooms or insufficient stale air getting out, so good circulation of that air should be ensured through special, approved 'holes' in fire doors. Obviously the air conditioning needs regular maintenance, regular cleaning and changing the filters and so on. In one building, the motorised valve on the main stale-air extract duct from the building had failed, with the valve locked shut so that stale air was not getting out as it should. Increasing the proportion of fresh to recycled air should help clear SBS symptoms. It is theoretically possible for cold viruses to get through filters and to be spread by a/c, but this would be impossible to prove, there being every likelihood of person-to-person spread in an open-plan office, or in a crowded train on a commuting journey.

If relative humidity (RH) falls below forty per cent in winter, the lining of the nose dries up and germs may gain access more readily. RH must be kept above this level – indoor plants which need regular watering help here.

LAVATORIES AND TOILET FACILITIES

Lavatories must be well lit, well ventilated and clean. Pressure of work and failing to supervise and question office cleaners (for instance, asking Have you used the lavatory brush?), can allow standards to fall to a level which can be hazardous to health. Cleaning lavatories is not a pleasant task and so may be skimped, particularly if not checked – a male supervisor may not check the ladies' loo and vice-versa.

Cleaning staff lavatories twice each day is ideal; regular inspections should take place. The lavatory flush and door handle can carry food-poisoning germs; a foot-pedal flush avoids some of this, and extraction flushing prevents the fine spray from conventional flushing if the lavatory lid is left up or if there is no lid.

There must be conveniently accessible, clean and well-lit washing facilities. Some employers also provide showers. Adhesive 'Now Wash Your Hands' notices should be placed on the inside of each lavatory door and are mandatory in catering unit toilets. Liquid or flake soap from a dispenser is better than a bar of soap, the undersurface of which becomes slimey.

Towels issued by dispenser, disposable paper towels or the pull-on linen continuous loop, are much better than the old-style roller towel which soon becomes dirty. A regular check is necessary to ensure adequate supply. Nothing gives a less caring impression, especially to people who have just washed their hands, than an empty paper-towel dispenser or the soiled end of a linen-roll towel dangling limply towards the floor.

Warm-air drying takes more time than most people will wait but can be usefully combined with paper towels.

A thoughtful extra is a solid-wick air freshener.

Repeated soiling outside the lavatory pan must be investigated at once and discussed at a staff meeting – not to be put off because of embarrassment. Soiling elsewhere in the toilets or in the building, for example in the lift, indicates a very disturbed mind and must be investigated at once. The person responsible needs urgent medical help.

Paint which cannot be written on, or ceiling-to-floor tiles, stop graffiti.

Water on the floor must be mopped up at once.

Enclosed jumbo size toilet paper rolls, locked in, prevent abuse and the paper getting spoiled – or stolen.

A drunk who is sick and makes an offensive mess in a staff lavatory, or elsewhere, should when sobered up be responsible for the cleaning-up, not leaving this for the office cleaners.

Ladies lavatories should contain means to dispose of used sanitary towels hygienically and safely, either into an enclosed, portable and free-standing unit containing chemical solution, with the unit being regularly changed, or into a wall-mounted plastic bag which is sealed at the end of the day and disposed of with the office refuse. Flushing non-disposable sanitary towels down the lavatory blocks drains.

As was mentioned earlier, in the section on Employment of Disabled, someone with paralysed legs (paraplegics) or paralysed arms and legs (tetraplegics) can also have urinary incontinence. This is satisfactorily controlled by a disposable bag attached to the leg, or by a highly absorbent pad within leakproof pants, and should create no difficulties with employment. A suitably adapted toilet nearby is necessary at the workplace. All modern office buildings should provide lavatory facilities for disabled people in wheelchairs, with sufficient door width and room to manoeuvre. A grab bar at the correct height helps get from wheelchair to lavatory seat, and back. There must be sufficient room for someone else to help if needed. The position and type of washhand basin, tap and soap dispenser, coat-hook and door lock require close attention so as to be usable by a person in a wheelchair either with or without full use of hands or arms.

CLOAKROOMS

There must be adequate cloakroom facilities for personal effects – coats, shoes, and so on – to be stored safely and securely. Ventilated lockers provide privacy. As these rooms can easily look scruffy, staff need an occasional reminder to keep them tidy.

PERSONAL HYGIENE

Personal hygiene is a delicate subject, but with people working close to each other, must not be glossed over. Harmony in an office is disrupted by claims that a member of staff 'smells'. Impeccable personal hygiene is even more necessary in catering departments to reduce the risk of food poisoning. This is discussed in the relevant chapter.

Some people do sweat more than others. A few perspire excess-ively – 'hyperhidrosis'. Of these, some are unaware, but usually the condition embarrasses them and stains their clothing. Medical treat-ment with a special aluminium based roll-on is usually effective, and avoids the need for the alternative – an operation to remove sweat-producing glands from the armpits.

Pure cotton or wool socks rather than nylon, and leather-soled shoes rather than synthetic, help sweaty feet; there are also dispos-able charcoal insoles to absorb perspiration and odour. Some people just don't wash or change their clothes often enough.

The need for direct confrontation can sometimes be avoided by occasionally including the subject at staff meetings and induction courses. Each new member of staff could be given a pamphlet on the subject, such as the one produced by the Health Education Auth-ority.

A talk can be given by the company nurse, which includes hygiene, with other health topics, and a description of the facilities provided.

No one person is singled out, but where a lady member of staff is suspected of poor hygiene, the nurse can be involved; if there is no nurse, and for all men, the supervisor or manager of the same sex should have an informal chat and give appropriate advice. One way of leading in to the subject is to say 'I have been advised that you and one other member of staff have a perspiration problem' – the other person does not exist.

Poor hygiene is usually more apparent in the summer and contributes to frayed tempers. A check should be made that ventilation systems are working properly and that a vendetta is not being waged by other staff.

Smells of all kinds can be disguised by an aerosol spray. The better quality sprays are oil-based and so stay in the air much longer than the cheaper water-based varieties which soon sink to the floor.

In a lavatory, a simple way to clear a smell is to light a match and then blow it out.

NOISE HAZARDS

Continuous or repeated exposure to very loud noise will damage hearing, and cause deafness sooner or later. Hearing loss is at first temporary, with recovery within a few hours away from the noise, but if no heed is taken and exposure to that noise continues, permanent hearing loss occurs. When this deafness begins, hearing loss is at first quite rapid, but the rate of progress then slows.

Very loud noise is most likely at, for example, road works, airports, garages, factories or dockyards. Office noise is much less, and such noise is more an unwanted sound and as such is a subjective assessment.

Noise has psychological, masking and physiological effects. The psychological effects are irritation and annoyance at what can be quite quiet nuisance sounds – people talking, a dripping tap, the squeaking hinges of an un-oiled office door. Such noise at work impairs efficiency and can contribute to a tension headache and fatigue. As the loudness increases, particularly if there are distinct, high pitched, interupted musical tones, irritation becomes annoyance and anger. Again there is much individual variation; what upsets one person may be hardly noticed by another. Masking noise blocks out sound-warning signals as well as normal conversation and so increases accident risk.

Physiological effects of very loud noise, apart from hearing loss, include faster metabolism with a rise in pulse rate, blood pressure and rate of breathing. Increased sweating of the hands is more linked to the psychological effect.

Sound is a form of vibratory energy – the collision of air molecules. It consists of pressure fluctuation waves, and the rate at which these occur is the frequency, measured in hertz (hz): a unit is one cycle per second. Both frequency and amplitude determine the loudness/intensity and the pitch. The audible frequency range for an adult is 15 hz to 20 000 hz – the latter figure falling with increasing age. Hearing loss is most apparent to the person concerned when it affects the upper end of the speech range at 4000 hz to 8000 hz.

Loudness/intensity is measured by electronic equipment (a sound-level meter), on a logarithmic scale in decibels (dB), filtered to simulate the human ear's particular sensitivity to some parts of the frequency range (63 hz to 63 000 hz), shown as dB(A). A logarithmic scale is used because of the very large span of sounds measured; the ear perceives sound in a particular way. For example, doubling the noise source only increases the sound measured by 3 dB(A), that is 90 dB(A) plus 90 dB(A) equals 93 dB(A). Examples of noise levels in dB(A) are:

jet engine 25 metres away: 140
jet at take-off 100 metres away: 125
pneumatic drill: 100
heavy goods vehicle: 90
street traffic: 85
car: 70
office: 65
conversation: 65
whisper: 30.

Noise rapidly decreases with distance from source; double the distance away and there is only one half of the sound intensity, that is, 95 dB(A) at three feet equals 92 dB(A) at six feet and 89 dB(A) at twelve feet. The doubling rule also applies to the length of exposure to the noise: 90 dB(A) for eight hours is the same as 93 dB(A) for four hours and 96 dB(A) for two hours.

Hearing is measured by an audiometer, preferably in a soundproof booth and only when the subject has not been exposed to loud noise for sixteen hours or more, to avoid the transient hearing loss of 'temporary threshold shift' referred to earlier. Some audiometers are operator controlled, others automatically increase the loudness of each frequency in turn, 500, 1000, 2000, 4000, 6000 hz, until the sound is heard, and record this on a graph. Hearing loss starts at 15 dB(A) loss and is complete by 82 dB(A) loss. Taking a detailed history, looking for past noise exposure and/or ear disease is essential, as is examination of the ears for wax. Irritating sounds in open-plan offices can to a certain extent be alleviated by good maintenance and attention to detail – oiling hinges, for example, and fitting self-closing doors that do not slam.

Computer terminals and printers can be isolated from the main office, or surrounded by sound-absorbing materials and soft furnishings. Typewriter manufacturers now offer machines with special sound-absorbing casing. Carpets are preferable to hard-floor surfaces which create echoes.

Carpets, curtains and cushions help to absorb sound as do acoustic tiles on ceilings and walls. When fluorescent tube fittings hum, the control gear needs attention. Careful design, installation and maintenance avoid some noises – gurgles and drips from the plumbing for example – and reduce equipment noise. Quite apart from the irritating noises in an office from people or equipment, noise can come in from outside. Sound is formed by vibration producing pressure waves and is reduced by putting objects in the path of these pressure waves: a brick wall, for example, is very effective and considerably reduces traffic noise. Some modern architectural design replaces much of the outer walls of office blocks with ceiling to floor 'picture windows' and these are far less effective at reducing sound. Checking for and filling gaps or cracks around the windows helps, as does thicker glass or sealed double glazing or – if traffic noise remains a problem – even triple glazing. (If this leads to condensation, it is important to keep the window frames clear of black mould to which some people can become allergic.)

In a building housing many small companies, noise from the offices of other companies coming through party walls can obviously be reduced by a positive response to a polite request to be quieter. If the noise continues unabated, complain to the local authority and strengthen the sound insulation of party walls.

The Department of Employment Code of Practice is that the maximum permissible noise level for occupational exposure should be 90 dB(A). A simple guide is that at this level you need to raise your voice to make yourself heard. The intention was to protect the majority, 99 per cent, from hearing loss induced by noise over a working lifetime, taken as eight hours a day, five days a week, for forty years. It has been suggested that in these conditions 11 per cent rather than the implied 1 per cent would suffer hearing loss. So a level lower than 90 dB(A) should be the aim. A Code of Practice can be used to support prosecution under HASWA by the Enforcement Authority.

Successful civil claims have been made against companies for causing hearing loss. In certain well-defined industrial situations, noise-induced hearing loss is a prescribed condition for which compensation is paid.

Hearing loss from noise is first at the 4000 hz frequency, above the speech range, so for a while the individual may not realise anything is wrong; not hearing bird sounds on rustling leaves, for example, is unnoticed. After a while, the noise at work that had seemed very loud becomes less noticeable. This is put down to having 'got used to the noise', whereas in reality insidious deafness is developing and subconsciously the individual starts to lip read.

Noise of course occurs away from work too. Examples are gunfire for clay pigeon shoots, rifle fire and amplified disco music. Continuing exposure to loud noise results sooner or later (depending on the dB(A) rating) in hearing loss spreading down into the range of speech frequencies, affecting first consonants and then vowels: A complicating factor is that hearing sensitivity decreases with age, (presbycusis.)

Regular servicing of office machinery and replacing worn parts helps control noise; where possible not using metal. Surround the source of noise with noise-absorbing materials: cladding or baffles can be used, or an airtight sound-insulating enclosure which achieves a reduction of twenty to thirty decibels.

The individual worker can be protected in several ways. A system of job rotation will limit the time spent at work in a noisy environment. Hearing protectors muffle the sound entering the ears, but are not popular because of the feeling of being cut off from colleagues: providing a choice encourages their use. Ear Plugs are made from a variety of materials – disposable, fine mineral fibre plugs, with or without an outer shaped and perforated polythene film; foam that is shaped by hand to fit the ear canal; or fixed shapes in plastic or rubber that must be kept clean. Using these devices cuts noise by up to 35 dB(A). Personalised ear muffs completely enclose the ears and fit snugly to the side of the head: noise is reduced by up to 40 dB(A). Music can be played into one earpiece and instructions given into the other. Ear defenders feel a little strange at first, but after a short while they are usually well tolerated provided they fit properly, and the employee gets into the habit of putting them on before entering the noisy environment.

Requirements in an EEC directive may spread to the UK in part or in whole by 1990 with hearing protectors recommended for anyone in the vicinity of 85 dB(A). If the noise is above 90 dB(A), the employer must try to reduce this at source; failing this, notices must warn of a noise-hazard area with signs to make everyone – employees and visitors – aware of it; all these persons must be provided with hearing protectors. A choice of three types of hearing protector must be made available where the noise level is above 85 dB(A), so that people may wear these if they wish. There will need to be training in the use of hearing defenders, and staff are to be made aware of the reason why they should wear defenders, which must also be properly maintained.

The employer is to keep a list of employees who may have to work within the identified noisy area and so be exposed to these noise levels. The employer must also keep detailed records of noise survey information, its date, who has access, who has been issued with ear protection and what type of ear protection, plus a chronological record of people entering and leaving the installation etc.

The employer is required to display and publish noise level readings for each piece of equipment and these noise level readings should be checked periodically to ensure that equipment wear does not result in an increase in noise level which might go unidentified.

VENTILATION

Effect on contact lens wearers

Staff who wear contact lenses may complain of sore eyes from working in an air-conditioned building. The surface of the contact lenses needs to be clean and moist, as does the surface of the eye. This is normally achieved by natural lubrication by tear flow and regular blinking, but air conditioning, especially in winter when the air is warmed, may dry the surface of the eye and of the contact lenses, and irritation follows. In addition, wearers of contact lenses are particularly susceptible to direct and reflected glare. There are many ways of reducing such glare by attention to lighting, work surfaces, the type of paper used, and so on.

Ionisation of the air

Air is mostly made up of oxygen, carbon dioxide and nitrogen molecules. Each molecule is made up of atoms. An atom has a central positively-charged nucleus around which move negatively-charged electrons. When the charges balance, the atom is electrically neutral. Remove electrons and there is a positive charge. Add electrons and there is a negative charge. Charged molecules, temporarily unbalanced, are positive or negative ions.

Ionisation is a natural process, produced for example by ultraviolet light, by lightning and by waterfalls. Artificial ionisation of the air, produced by a high-voltage generator, is claimed to be of benefit in several medical disorders including respiratory conditions like asthma, hay fever, bronchitis, sinusitis and catarrh; skin burns are said to heal faster and without secondary infection; anxiety responses are lessened, heart rate and blood pressure fall, and breathing capacity increases. It is claimed that these changes are not merely placebo effects and that it is negatively-ionised air that has a beneficial effects. Apart from specific effects, subjects apparently feel better generally, get less tired during the day, and have an increased sense of well-being.

Metal-ducted air-conditioning systems, especially those using re-cycled air, are said to reduce negative ionisation, as does air pollution, be it from car exhausts, chimney smoke, cigarette smoke or dust. Particles in the air 'mop up' the negative ions, which lose their effect, while static attracts negative ions. An office in a built-up area thus has a preponderance of positive ions. There is no consistent, clear, authoritative medical evidence this matters a jot, but apparently in one trial, equal numbers of negative-ionising machines and 'dummy' machines were put for four weeks on the desks of fifty single occupancy offices and the occupants asked to continue the daily assessment of how they felt, something they had already been doing for a month. Those with the negative ionisation machines reported a significantly better sense of well-being; the atmosphere seemed fresher, headaches were fewer and there were no adverse effects. On the other hand, a similar trial in the City showed no benefit what-soever. Thus, whilst anecdotes abound, there is no consistent or objective evidence of benefits. Ionisers do however convincingly clear an area of cigarette smoke, and so may have a place in a mixed office of smokers and non-smokers. Smoke particles once given a negative change and attracted to the floor which has a positive change. Likewise an ioniser traps fine dust.

Air-conditioning hazards

In some air-conditioned offices where the system uses a resevoir of recycled water for humidification, humidifier fever has occurred, caused by sensitivity to products of germs or amoebae which get into the system in the air or the water used, grow on the humidifier baffle plates (designed to eliminate droplets) and are nourished by organic dust which enters the system through or round air filters. These products are then blown out through the ducts and vents into the offices. Not everyone is necessarily affected after repeated exposure, those who are, get flu-like symptoms soon after starting work after a few days' break, for example after the weekend or a holiday – hence the term 'Monday morning sickness'. Symptoms include fever, fatigue, aching all over, a cough, and pain tightness in the chest. Within a day and on no treatment, the person feels fine and can work again, but the next Monday the same problem recurs. How many are affected depends on the degree of contamination. When this is heavy almost everyone gets it and when light only those who are susceptible to allergies. The allergic factor is confirmed by skin-testing.

Humidifier fever is commoner in the winter and is similar to farmer's lung, caused by moving mouldy hay and, as in this condition, if allowed to go on and on, can cause permanent lung damage. Good housekeeping of the air-conditioning system prevents humidifier fever as well as legionnaires' disease (see below). This means regular inspection, cleaning and maintenance of the humidification unit and of the whole air-conditioning plant including the water cooling tower, getting rid of all sticky dirt, grease and rust, adequate, properly fitted filters on air intakes, especially into the humidification system, and the use of chlorination and/or specific biocides against Legionella.

Germs may also grow in the dust that collects at the bends of air-conditioning ducts which should be cleaned when fitted in the first place. These ducts can be forgotten – 'out of sight, out of mind' – and may need maintaining. Portable office spray humidifiers can also become contaminated.

The word 'legionnaire' conjures up dramatic scenes of the French Foreign Legion marching through the desert, but nowadays this word is more associated with a waterborne germ that in certain circumstances gets into the air, surviving up to two hours, and can be inhaled and cause illness. The germ is Legionella Pneumophila – the illness is legionnaire's disease.

This germ was first identified after causing the deaths of twenty-nine veterans at a Legion Convention in a hotel in Philadelphia, USA, in 1976. Outbreaks have occurred throughout the world, mainly in hospitals and hotels. Out of about a hundred and fifty thousand cases of pneumonia each year in England and Wales, legionnaires's disease is responsible for about two hundred. The germ is very common in and around natural water, in ponds, rivers and streams and in the mud on their banks. In this form the germ is harmless. Only when it is in a droplet aerosol spray is there a danger.

The bacteria grow best, and that can be very fast, in a warm environment: for the legionella this means warm, stagnant water at between 20 and 45 degrees centigrade. Most outbreaks occur at the end of the summer. Some have been in holiday hotels abroad, where the stagnant water has been in the en-suite showerhead. The first person to take a shower gets the contaminated spray. Large calorifiers and 'dead leg' pipework are also sources.

The other common source are the water-cooling towers of air-conditioning systems on the tops of modern office blocks. Here warm water is cooled by air being drawn up through a falling spray of the water. Warm water can allow legionella to breed, and despite drift eliminator slats, aerosol droplets may escape, into the discharging air and on to people passing by below; if the fresh air intake of the air-conditioning unit is close by, the droplets are also drawn back into the building itself.

Where air-conditioned buildings have water-cooling towers, there must be a comprehensive procedure for regular maintenance and cleaning of the system, to eliminate sludge, scale, rust and algae which encourage the legionella to breed. Cooling towers are disinfected by chlorination and/or biocides, and the water system should be designed to avoid stagnation.

After three to six days' incubation, legionnaire's disease starts like a bad bout of influenza, with a fever, aching muscles, headache and a dry cough. Influenza is caused by a virus and clears in a week or two, but the legionella bacterium can go on to cause pneumonia, for which a specific antibiotic is needed for cure. There can be other symptoms including diarrhoea and confusion. 10 per cent of those affected die, but the condition may be so mild as to be virtually unnoticed.

Another form is known as Pontiac fever. Legionnaire's disease must always be considered by doctors in the differential diagnosis of atypical pneumonia. Those aged between forty and seventy are particularly at risk. Three times as many men as women are likely to be affected. Healthy people may develop legionnaire's disease, but smokers, alcoholics, and patients with cancer, diabetes or chronic lung or kidney disease are particularly at risk.

Legionnaire's disease, unlike influenza, is not infectious, it cannot be passed from person to person. Because legionella is widespread, sampling water systems may identify the germ, but good housekeeping and a well designed system will prevent legionnaire's disease. Growth of legionella in domestic water supplies can be prevented by keeping cold water sufficiently cold, below 20° centigrade, and hot water sufficiently hot, above50° centigrade. The hot-water cylinder should also be checked, as legionella can lurk at the bottom. Rubber washers are being phased out, as legionella grows on these, and replaced by a composite non-rubber washer such as that produced by the Water Research Council.

Smoking at work

Whilst on the subject of ventilation and clean air, it is appropriate to discuss smoking at work and pressure from staff for 'no smoking' work areas. Nowadays 'No Smoking' signs are commonplace in public areas. For many years British Rail has provided 'no smoking' compartments; smoking is not permitted by London Transport anywhere on the underground; on double-decker buses, smoking is only permitted on part of the upper deck. 'No Smoking' areas have been created in restaurants; some cinemas, theatres and lecture halls prohibit smoking altogether. Most companies forbid smoking in rooms containing computer equipment (just one smoke particle on a floppy disc can make it 'go down', losing the entire contents); in food preparation areas (a legal requirement); staff restaurants where there is a separate coffee area; staff transport vehicles; lifts; waste-paper stores, and near oxygen, petrol or other flammables.

Numbers of adult smokers are dwindling and now over half of all adults do not smoke, but whilst fewer men smoke, more young women are doing so. 15 per cent of men smoke pipes and cigars. On average, males smoke more than twenty cigarettes a day and female smokers less than twenty. More men inhale than women. Cigarette smoking is habit-forming and addictive. (Nicotine is said to be more addictive than heroin.) Many smokers have tried to give up.

Some companies in the UK and the USA have a policy whereby, if the majority in any office do not smoke, they can choose to designate the office a 'no smoking' area for all staff and visitors. 'Thank you for not smoking' notices are provided.

It may be possible to segregate smokers from nonsmokers in large organisations where many people do the same type of work.

Laws have been passed in parts of the US giving a common-law right to a smoke-free environment at work, or banning smoking in virtually all work places and public places, and imposing restrictions on smoking in restaurants and private offices. In New York, for example, an office worker can declare his or her work-station, and eight feet around it, smoke free.

Depriving smokers of the opportunity to smoke interferes with freedom of choice, but a smoke-filled atmosphere is not only irritant to the throat, nose and eyes, but has been shown to increase the risk of illness in nonsmokers from this passive, 'forced' smoking. So far, evidence is based on nonsmokers who live with smokers, and the extent of the link with the workplace alone is not as clear. Side-stream, unfiltered smoke, curling up from a lit cigarette on an ashtray, is especially harmful.

Cigarette smoke contains many potentially hazardous substances, including nitrosamines, benzopyrene and carbon monoxide. Carbon monoxide, for example, enters the blood stream in preference to oxygen. Especially those with chronic lung or heart disease need all the oxygen they can get, and may suffer from breathing in other people's smoke. Children whose parents smoke have more respiratory tract infections than children of non-smokers. Overseas, a successful claim has been made on the employer by one man whose asthma was brought on by working in a smoky atmosphere and by another whose bronchitis was made worse. Most people now realise there can be risks to health from cigarette smoking.

Local concensus can be more acceptable than a central directive: to add the proviso that where there is deadlock, the wishes of the nonsmokers should prevail, is in effect a no smoking policy.

No smoking may be a requirement of employment, as in one company where the managing director made this ruling when the company moved to new premises. It was partly because of the effect of smoking on décor in a newly-decorated building, and it only needs

a glance at the walls to see the difference between the clean restaurants where smoking is not permitted, and the grimy walls, curtains, furnishings, and so on in the coffee areas where smoking is
allowed. No-smoking may be introduced in a voluntary manner, by
staff who do not smoke displaying positive 'Thank you for not
smoking here' notices. Staff operating visual display units may complain of eye irritation, realise this is brought on or aggravated by
cigarette smoke, and thus avoid smoking. Well-ventilated rest areas
could be provided where smokers can smoke in break-times.

Sir Richard Doll, the scientist who established the link between
smoking and lung cancer in the 1950s, said in 1986 that 'some cancers
must certainly be caused by the involuntary or passive inspiration of
smoke that others produce.' The risk to nonsmokers has been estimated at between 0.5 and 10 per cent of the risk to smokers themselves.

The organisation ASH – Action on Smoking and Health – has
sponsored seminars about smoking at work and has produced a
booklet, 'Action on Smoking at Work'.

LIGHTING

Eyestrain, i.e. actual harm to the eyes from using them, does not
happen. In other words, however long close work is continued, the
eyes are not damaged; but with unsatisfactory environmental conditions – the wrong lighting, too dim, too bright, glaring or haphazard,
with lighting levels perhaps based on domestic experience, eye discomfort develops and so performance is impaired. Also, in a dimly-lit
area there is more risk of tripping over or knocking into something,
to say nothing of making clerical errors.

Several organisations and government departments have lighting
standards to meet their particular requirements, but there is little
precise legislation. Guidelines are contained in the standard reference for lighting engineers, 'The Illuminating Engineering Society
(IES) Code for Interior Lighting'.

Light output is measured in 'lumens' and the illumination produced is measured in 'lux'. To give some idea of figures, outside on a
bright, but not sunny day, there are 5000 lux; step back inside a
picture window and the level is halved; 'bad light stops play' at 100
lux.

Office lighting has been found as low as 200 lux. The efficiency of the eye improves with more light, up to daylight levels. There are, however, sharp, differences in what people prefer. A majority of office staff like to work at 500 lux, in line with IES recommendations. The rest are split between 1000 lux and less than 400 lux.

A suggested scheme for minimum levels of lighting for offices is as follows:

Reception areas and corridors	200 lux
Managerial rooms	500 lux
General offices	500 lux
Typing and machine areas	750 lux
Filing areas	300 lux

These levels can be achieved by either general or local lighting. Providing a desk 'task' lamp for each person helps when the general lighting level is kept down to avoid, for example, reflections on VDU screens or glare but compromise should suffice. The colour, brightness and shininess of walls, floors and ceilings affect user comfort from lighting levels.

Eyestrain, better called eye fatigue, is made worse by flicker or glare. I find it amazing how a faulty fluorescent tube's flicker is tolerated, even unnoticed, when all that is usually needed is to change the tube. Susceptibility to glare, 'the subjective experience of large deviations from the overall level of lighting', varies widely. Glare can be tackled by modifying existing light-fittings, raising them, improving the diffusers or adding shades. Turning the light source upside down, that is, bouncing light off the ceiling ('up lighting'), is wasteful in terms of electricity used, but prevents glare. Get rid of highly-reflective surfaces. Glare from horizontal bright lights can be direct or reflected from glass, gloss paint, polished desks, cupped and shiny typewriter and computer keyboard keys, white shiny paper and so on. Lights can be shaded, windows covered with adjustable blinds, paintwork a matt surface and working paper matt; pastel – yellow or green – are good paper colours.

The eye is not strained by use, but the ciliary muscle which focuses the lens can be fatigued. This fatigue makes the eyes feel heavy and uncomfortable. To counter this, it helps to pause for a few moments now and then, to look into the distance and allow this muscle to relax. Such symptoms may also be due to an existing defect of one or both eyes of which the individual may have been unaware. Up to a third of staff of all ages are said to have less than perfect vision. Anyone at all concerned and everyone over age forty should go to an

NHS ophthalmic optician or optometrist for an eye test every two years. A quick preliminary check is by reading a car number plate twenty-five yards away and the page of a book from about nineteen inches.

There is no authoritative evidence that a basement or windowless environment without daylight harms health, but curioulsy, physical strength may be less – there is no known explanation for this. A blank wall close by limits eye-muscle relaxation, and one novel solution is to display a panoramic picture all along one wall and even put this behind fake windows, to represent a distant view.

Some staff are not happy away from natural light; changing to warm white fluorescent tubes or total spectrum fluorescent light may help. High frequency fluorescent tubes reduced the incidence of headaches by fifty per cent in staff working under them.

Linked to the amount of daylight is the newly-described Seasonal Affective Disorder (SAD), a depressed mood which comes on in winter when there is least daylight. It is said to be helped by exposure to total spectrum light, but there is no conclusive evidence the condition even exists: winter can be a gloomy time for everyone. There is no real support for the claim that fluorescent lighting induces skin cancer; newspaper headlines on this subject caused understand-able alarm. Certain drugs make the skin sensitive to ultraviolet light, but the amount of ultraviolet light reaching the skin from fluorescent tubes at ceiling level is so small as to have no effect whatsoever in my view; the same is true of the output from a VDU screen.

Working environment check list

Date/Time
Weather/Outside Temperature
  (1) *Air*

> Room Temperature. . . . . C (NR. 19–23 C)

> Wall Temperature no more than 3 C above or below air temperature; no surface temperature over 50 C.

> Relative humidity. . . . (NR. 40–60%)

> Air movement seated
> head. . . . mm per second

Knee. . . . mm per second (ideal 200mm per second –
draught 500 mm per second).

Air changes at 1.3 litres per second per metre.

(2) *Noise*

Floor covering . . . Carpets/lino/wood/stone/other.

Continous loud noise?

Move the source elsewhere, or use an acoustic hood.

Intermittent loud noise?

Move the source elsewhere, or use an acoustic hood.

Maximum continous noise. . . . dB(A)

If telephone in use, aim for under 55 dB(A); if more than 60
dB(A), consider acoustic treatment.

(3) *Light*

Ceiling height. . . . . metres

Tasks:

VDU/Clerical/VDU and Clerical

VDU ideal 300–500 lux.

Clerical ideal 500–750 lux.

Minimum, lights on, furthest from windows, blinds shut. . . . . lux

Minimum, lights off, furthest from windows, blinds shut. . . . . . lux

Maximum, lights on, by windows, blinds open. . . . lux

Maximum, lights off, by windows, blinds open. . . . . lux

Desk lamp and document holder may help VDU operators
read source documents.

Room lighting?

Fluorescent tubes with diffusers parallel to VDU operators'
line of sight and parallel to windows.

Which walls have windows; which direction(s) do the windows
face?

What blinds are fitted?

How is lighting controlled?

Is the VDU screen treated to prevent reflected glare? If so, how?

Is a VDU filter used? If so, what type?

Is legibility reduced? Reflections?

Ceiling reflectance? (ideal 80–90%). Walls? (ideal 30–80%) (no gloss paint).

Floor? (ideal 20–40%)

SHIFT WORK AND FLEXITIME

Some computer operations and other workstations must be manned round the clock, and in other situations there can be cost savings in keeping machinery going for twenty-four hours a day. This means shift work. Some people claim that shift work, particularly where this includes periodic night shifts, damages health. This view is based on subjective opinion and the experience of people who cannot settle into shift work. Research shows that shift workers have less sickness absence than those on permanent days, both sickness absence with a statement from a doctor, that is, eight days or more, and self-certificated sickness absence, that is, seven days or less, as well as less absenteeism. The reason for this could be that shift workers usually work in small groups: morale is high and the absence of one person would jeopardise the efficiency of the group; also, there is often less direct supervision at night. Shift work is usually voluntary; the self-elected groups are therefore unrepresentative of the whole working population. Imposition of shift work would be unlikely to show a decline in sickness absence figures.

There are a few medical contraindications to round-the-clock varying shift work with regular alteration in the hours worked. One example is insulin-dependent diabetes, where it is best to have a set routine of injections and meals; another is the type of epilepsy where fits only occur at night.

There are, of course, social disadvantages to some types of shift work, particularly the night shift, but a report by the Prices and Incomes Board showed these are often matched by advantages; apart from having more money, some people welcome the variety, though there would be discord if the spouse is unhappy or resents being left alone at home at night.

A variation of shift work is 'flexitime', flexible working hours. The employee chooses, within limits, the times of starting and finishing work, but has to work a central 'core', commonly 10 a.m. to 4 p.m. On a monthly cycle, once the basic normal working hours have been completed, extra hours worked are logged and go towards additional days of annual leave. Sick leave is separate from flexitime, but time off for a dentist or optician appointment, previously 'given' by the employer, is now included in the flexitime total with the obligation to make it up. The work of many companies cannot adapt to fit the flexitime system.

11 Work Surroundings and Equipment

FLOOR-SPACE REQUIREMENTS

The law says there must not be overcrowding by people, furniture or machinery so as to cause risk of injury to health. Building design and floor weight or local regulations may restrict the numbers working within a certain area.

The minimum, permitted floor space for each person depends on ceiling height. Normally forty square feet is required per person, but where the ceiling height is lower than ten feet, there must be 400 cubic feet per person. These requirements do not include space occupied by fixtures and fittings. Employing three people in a room with a ceiling height of eight feet means at least 1200 cubic feet of floor space is necessary, that is, a room area of one hundred and fifty square feet – fifty square feet for each person.

These standards do not apply to areas such as shops where members of the public will be, though here, too, under normal conditions overcrowding must not occur; an obvious exception is the first day of the sale at a major department store.

With regard to the air, in all work rooms there must be an effective and suitable means of ventilation by the circulation of adequate supplies of either fresh or artificially-purified air.

LAYOUT AND DECOR

Good layout and decor of the work place are important for staff morale. High office rents make it tempting to cram in as much as possible, but as stated, the load the floor can take must be considered and heavy items spread out. Congestion and clutter cause accidents. There is also a legal requirement to provide an adequately defined passageway to exits.

Trailing telephone and other wires and cables are another accident risk. Wire management is an important consideration in workstation design, with manufacturers cleverly tucking wires away along the edges of desks and partitions, yet with ready access. In new buildings,

160

the power supply comes through recessed boxes in the floor under carpet tiles.

A large, open-plan office can be a daunting place to work, inducing the feeling of being a number rather than a person. Where possible, the area should be broken up with screens which give a degree of privacy and sound insulation and allow some personalisation, as a place for displaying photographs and postcards, for example. An alternative is the use of partitions as partial room dividers.

Provided they are looked after, evergreen plants or fresh flowers soften a work room. The choice of plants depends on where they are to be sited. For example, some plants do not mind draughts whilst others wilt and give the place a sick-building look. Plants, regularly watered to maintain a good level of natural humidity, counteract the drying effect of electrical equipment on the atmosphere – something not all air-conditioning systems can cope with as the amount of that equipment increases. Maintenance-free plastic plants do not have that advantage.

Background music can be helpful where there would otherwise be silence, but some people cannot stand it: where the facility exists, the majority in each office or department should decide whether to have music at all, and if so, whether to have it only at certain times of the day.

Colour schemes also provoke strong feelings. The use of the right colour scheme helps to create a calm yet vibrant atmosphere. Different colours are associated with definite feelings, moods and concepts.

The primary colours: red, yellow and blue, are augmented by green, orange and purple, the secondary colours. Adding white to any colour creates a tint and lightens; adding black creates a shade and darkens. Blue recedes, gives an impression of space, is associated with coolness, is pacifying and relaxing. Red is associated with warmth, it advances and is aggressive. Yellow, the sunshine colour, is 'active', whereas black, the colour of night, is 'inactive' and passive.

Tints and shades have the same effects as the parent colours but are more subtle. Neutral grey or semi-neutral mushroom are colours with which all other colours blend. Where maximum light is needed, pale colours are used to reflect all available light, though plain white gives a harsh, clinical effect and contributes to glare.

In the communal parts of a large building, single bold colours, on one wall, or on pillars, are successfully used to identify different areas. On the continent, bold colours are increasingly used in offices in place of more neutral colours, making for a bright, optimistic atmosphere.

An example of the effect of colour is that of a company moving to a new building where soon most of the staff complained of tiredness, irritability and eye strain. Management concluded the lights were wrong and asked for them to be changed. The lighting engineer privately considered the lighting to be perfectly satisfactory and decided the décor was at fault – all the walls were pale grey. At the weekend his work force painted the office in bold colours – the lights were not touched. A week later he was told 'Everyone now feels fine. Thank you for dealing with the lights – and you painted the place as well!'

CHAIRS AND DESKS

To give of their best, staff need comfortable chairs. The chair back must support the forward lumbar curve in the lower back, and the seat needs a sloping, soft, 'waterfall' front edge, so that the thighs and the backs of the knees are not under pressure. Such pressure could obstruct blood circulation or press on nerves; indeed this edge should be six to eight inches away from the back of the knees. If it is further forward, staff will perch themselves on the front edge and not use the back support at all. Staff may not realise their chairs can be adjusted, and may think of them as like dining-room chairs at home, or they may not bother to adjust them, or may not be able to work out how to adjustment them, or have lost the manufacturer's guid-ance notes (or never found them in the first place, as so often these notes are attached to the underside of the chair). The end result is that the backrest and/or the height of many office chairs are incor-rectly adjusted for whoever is sitting on them, or are incorrectly adjusted for the work surface, and so fail to give adequate support.

At one talk on chairs, a designer discussed the use of deep, square upholstered chairs without arms, intended to discourage sitting for more than a few minutes and suitable for a foyer! Another chair rated uncomfortable was said to be suitable for boardrooms! Some chairs have a straight back support. A typist's chair may have a narrow vinyl-covered, padded, adjustable bar that should support the lower i.e. lumbar spine, but may have been wrongly adjusted and presses into the wrong place. There is better awareness that we need firm support for the lower spine, and where this is missing, an orthopedically-designed foam wedge can be placed in the angle between the seat and the back rest. Hopefully gone from every office is the executive-type chair which swivels, reclines and is well padded, but the back curves in the wrong direction, concave rather than convex above the seat, providing no support. Sitting slumped in one of these chairs stretches ligaments and contributes to backache.

Another common fault is a seat which is too long, as mentioned earlier, so that it is impossible to have the back supported and the feet on the floor. To get their feet on the floor, the person then has to perch on the front on the chair, well away from the back support. If you look round any office, however well designed the chairs are, you will see many people leaning forward, not using the back support at all.

The best standard office chairs now available will be suitable for a person susceptible to backache, and avoid the need for an ortho-paedic chair. The Balans principal – of a sloping seat without any back support and a kneeling bar – has not found favour in British offices in my experience. An aid which partly produces this effect and helps a person with low back pain is the Gibbs wedge placed on the seat of a conventional office chair.

There are British Standard Specifications and measurements for the ergonomic and functional requirements of office furniture: the chair, footrest, worktop, drawers and so on are recommended to be considered as a matched unit. The Health and Safety Executive (HSE) also makes recommendations about the dimensions of office furniture. The intention is to ensure that the worker's arms are free and unrestricted; the eyes a comfortable distance from the work; the feet resting easily; the lower trunk properly supported, with adequate space to fidget, that is, to make the very necessary changes of posture needed to avoid muscle stiffness and fatigue.

The range of adult dimensions is immense, but it has been found that the specifications for fixed work surfaces and adjustable chairs will suit at least ninety per cent of the population.

The HSE, in the HMSO publication 'Seats for Workers in Fac-tories, Offices and Shops', points out that many office desks and chairs are too high for comfort, and that footrests, unpopular and difficult to keep clean, tend to disappear. The HSE recommends the desk be not more than twenty-eight inches high and the kneehole (without a drawer) twenty-six inches, used with a chair with a seat seventeen inches high to enable the elbows to be level with the desktop.

For typing, twenty-five inch surface and twenty-four inch kneehole are recommended to be used with the seventeen inch chair. Where work has to be done standing, thirty-eight inch high tables are recommended.

There are also general design requirements to look out for when choosing office furniture. Edges, corners and protrusions should be rounded and smooth to avoid injury or damage to clothes. Rounded corners for office workstations cost more to manufacture. A neat, cheap solution to 'round off' sharp corners is a plastic device held on with a sticky pad which completely covers the corner. All office furniture must be stable, that is, it cannot be readily tipped over, and of low fire risk; electrically-safe where applicable, and with any lubricated parts free from leaks. Parts which are moveable and adjustable must be safe and secure. The typical office desk is far from perfect. Usually too large, its right-angle edges and corners are painful to bump into. Trailing wires from telephones, an electric typewriter, a computer and/or dictating machine, drape themselves over the side, entangling passersby and causing trips and falls. Leaning over a flat surface can strain the neck; better, but rarely seen outside the offices of designers and draughtsmen and women, are workstations on the lines of school desks, with a sloping work surface. A simple solution, and worth a try, is to slope a hardbacked blotter.

Accidents to office staff can occur from leaning back in a chair and overbalancing. If the chair was defective, it should have been reported and repaired, and both the employer and the employee, if aware of the fault, are liable under HASWA. The employer must ensure office furniture is in a good state of repair. An accident can also occur if a swivel chair is tipped back on its castors and topples over; a swivel chair with five feet supporting the central pillar is more stable than one with four.

Very occasionally there has been a report of a gas-lift cylinder exploding. More chairs now have this method of height/tilt adjustment and the risk of explosion is confined to side entry of the adjustment lever – entry on top of the cylinder is perfectly safe. No one weighing over sixteen stone (100 kg.) should sit on a gas-lift cylinder chair.

VISUAL DISPLAY UNITS AND WORD-PROCESSING MACHINES

Visual Display Units (VDUs), also called Visual Display Terminals (VDTs), are used more and more in offices to gain access to and then interact with computer systems. Two screens and two keyboards per work station are not uncommon, and some offices have banks of VDUs lined up side by side. Indeed, it can prove impossible to avoid being in the vicinity of a VDU. Most VDUs use cathode rays on the same principle as television sets, but to display letters and figures instead of pictures on a screen; below is a separate keyboard for passwords, enquiry and data entry.

There is a world of difference between watching a favourite TV programme at home and searching for information on a VDU screen at work. In the years VDUs have been in use, there is no proof of anyone using a screen being made seriously ill as a result, but in recent years there has been great concern expressed by employees, and by the unions which represent them, that VDUs (like smoking) can seriously damage health. What can get overlooked is the great boon provided by the new technology revolution, making the search for and retrieval of information much easier, and removing the drudgery of retyping innumerable drafts and standard letters.

Way back in 1975, a report from France said that trainee VDU operators should be under forty-five and receive instruction for no more than four hours a day. Initially omitted from the English translation was that the subject of the report was training. It has been claimed that a VDU is hazardous to sight, just as very loud noise is to hearing. Little wonder that some VDU operators become worried. The principal areas of concern initially were ionising radiation and eyestrain. Later, possible effects on pregnancy and upper limb disorders have been of concern. Eyestrain suggests actual damage to the eyes from over-use, but as mentioned earlier, this does not happen. The eyes are remarkably efficient and resilient and cannot be damaged in this way; likewise, the loss of one eye through injury or disease places no extra strain on other eye. The eyes are not damaged by glasses of the wrong prescription nor by poor lighting. In middle age, the lens of the eye become less elastic (presbyopia) and so close work in general, and, for the office worker, reading in particular, gets more difficult; sooner or later, reading glasses are needed. This is now happening rather earlier in the Western world – in the early rather than the late forties. It needs emphasising that there is no evidence that working with a VDU damages the eyes or accelerates the normal ageing process.

Concentrating on any close work, including prolonged VDU work, is tiring and good eyesight is obviously necessary. A VDU operator may be constantly refocusing on objects at three distances and three angles from the eyes, namely the screen, which will be near-vertical, the keyboard, which is inclined, and horizontal input material flat on the desk, be this copy letters, computer printout or other papers.

Eyestrain, headaches, discomfort and bleariness around the eye is better called eye fatigue, and is caused by unrelieved tension in the ciliary muscles which move and focus the eyes. Liability to eye fatigue is eased by momentarily looking into the distance or closing the eyes, allowing these ciliary muscles to relax. Also, regular blinking lubricates the surface of the eye. Eye fatigue, headaches and tiredness are more likely when the VDU is poorly adjusted, or the work station is not set up properly, for example, a migraine type headache can be triggered in a susceptible person by a flickering or over-bright screen image or by bright reflections on the screen. Contact-lens wearers have a particular problem, as the surface of the lens becomes dry sooner than the surface of the eye; concentration leads to staring and decreases blink rate, and this hastens surface drying. Eye symptoms result.

Tobacco smoke irritates the eyes: nonsmoking areas are therefore advisable. (Cigarette smoke can also damage floppy discs) The fact that existing defects of eyesight only become apparent when VDU work is started leads to the VDU being wrongly blamed as the cause of such faulty vision. As previously stated, anyone at all concerned, including prospective VDU operators, should have regular NHS eyesight testing anyway, if necessary in company time, to detect and correct imperfect sight. To insist on frequent repeat eye tests, some taking place as often as every six months, far from reassuring VDU operators, could suggest to them that VDUs harm the eyes: 'Why else would they check my eyes so often?' Where large numbers of VDU operators are employed, a machine, operated if necessary by a lay person, can provide a rapid preliminary screening test. As mentioned, at any age, if eyesight seems less than perfect, it is logical to visit an ophthalmic optician or optometrist so that eyesight can be properly tested. There is now no longer the need to first obtain a GPs confirmation that eyesight testing is necessary. Someone who normally has reading glasses may need a special pair of glasses for VDU work, where the screen is beyond normal reading distance, though being able to move the modern screen nearer should avoid this. It needs emphasising to the work force that there is no evidence that VDUs damage eyesight in any way. Also available for the wearer of bifocals, who needs to tilt the head back to view the screens and so risks neck pain, is a lens which gradually alters without definite lines and has a large area for the intermediate screen distance, of 400 to 600 mm from the eyes. With regard to the workstation, there is a need to distinguish full-time VDU operators from those who use a VDU in their work, or refer to it periodically, but are not looking at the screen all day long. Any VDU should be set up in the best possible conditions; those in constant use need special consideration (see the environmental check list as shown in this book).

Bad posture aggravates eye fatigue, so it is important the operator sits both comfortably and correctly in a chair which gives support to the lower back and thighs. The screen angle should be about sixty degrees to the horizontal for a 600 mm viewing distance: This provides comfortable angle of gaze, and minimises postural fatigue of the neck. The height of the keyboard and the chair seat should be adjustable. With the operator correctly seated, the upper arms should be nearly vertical, the forearms nearly horizontal, and the hands tilted slightly upwards so as to allow free finger travel over the middle of the keyboard without much arm movement. The computer keyboard should be adjusted so that the forearms can rest flat on the desk surface or in fitted wrist-rests, and the chair seat so that feet are comfortably flat on the floor or a foot rest.

The brightness, focus and background of the screen are adjustable to personal preference. Operators must know where the adjustment controls are. The screen is designed not to flicker and must be cleaned regularly at least once a day with an anti-static cloth or spray. Also not to be overlooked is the need for clear printing on printouts.

A very few people with epilepsy may suffer a fit from the flicker of a VDU screen – this is known as photo-sensitive epilepsy. This usually first happens in the teens, brought on by watching television, but it can happen from a VDU even though the refresh rate is now twice as rapid as that of a domestic television set. Reasons for this include sitting close to the set and viewing from the side. Photo-sensitive epilepsy accounts for only five per cent of the total number of epileptics, so most people with epilepsy are at no risk from operating a VDU.

The eye is most sensitive to the yellow-green part of the colour spectrum. This is probably why green characters on a dark background was the colour most favoured by manufacturers, but this can leave a few people with a temporary pink after-image – they see pale-coloured objects tinged with pink or edged in pink. This is least noticeable if the décor, colour of walls and so on are complementary rather than, for instance, plain white, which would accentuate the after-image. This 'McCullough effect', alarmingly referred to as 'pink eye', is physiological not pathological; it is disconcerting but does no harm, contrary to what one person said when unconvinced by this reassurance – 'how can you guarantee it is safe'. What has happened is that the green-sensitive pigment in the retina has been temporarily used up and so the person sees the complementary colour pink until this is restored. The latest screens are black on white and so avoid this.

There is less chance of a glare from a matt wall surface than from a gloss or silk finish. In many American VDU rooms, the walls are light-coloured, for instance cream or grey, and features such as columns are coloured more strongly. A bluish tone on low ceilings increases the apparent height. Otherwise blue is to be avoided unless the room gets warm. Scripts should be on matt pastel-coloured paper – yellow is found to be the best colour.

Adjustment of the vertical tilt of the screen can reduce reflected glare from ceiling lights. If this does not work put a filter in front of the screen, held on with velcro pads. A neater method, which avoids the need for rolls of sticky tape if the pads slip, is to put the filter behind the VDU casing. Some new screens have an etched surface on which reflections do not occur. The correct type and level of ambient lighting ensures adequate illumination for associated tasks, whilst avoiding glare. When natural light is the main source, the screen is best placed at right-angles to, never facing or backing on to, the windows. Blinds or curtains may be necessary in any case, if the sun streams in at some part of the day, to avoid the extreme variation in brightness of natural light. Undue contrast between the VDU and its surroundings should be avoided. For some of the day, in winter especially, the main lighting will be from overhead fluorescent tubes. Here the ideal is to place the workstation between rows of such lights so the operator sees them end on. Sometimes this cannot be achieved and then attention should be directed at providing the best diffusers. In some instances, glare is only reduced by switching off alternate overhead fluorescent tubes. 'Uplighting' (bouncing light off the ceiling) is an inefficient use of electricity, but it avoids this source of glare.

The heat produced by many VDUs and printers in one office is significant, and may raise the temperature of the room if the air-conditioning was not designed to cope with this extra load. Supplementary, user-operated, wall-mounted room coolers are effective and much better than fans.

Another fear has been of exposure to radiation, both ionising such as X-rays, and non-ionising such as infrared, ultraviolet and micro-wave radiation, but the output is so minute that there is no risk to health, and a VDU does not add to the normal atmospheric radiation level, indeed there is more radiation person to person than from a VDU.

Attention later turned to less well-known forms of radiation, very low frequency and extra low frequency, with concern about the possible effects on pregnancy from pulsed waves of a certain frequency and duration, above and below which this 'window' effect did not occur, but there is no confirmed medical evidence to support this theory. The National Radiological Protection Board (NRPB) has stated that the radiation from VDUs cannot cause physical injury to people working with them. Tiny amounts of radiation are given out from the back and sides of a VDU, but there is no indication for enclosing the VDU to prevent this, as even sitting surrounded by banks of VDUs, total radiation remains well below the threshold for harm of any kind. Another concern was exposure to polychlorinated biphenyls (pcbs) which cause cancer, but in the VDU these were sealed in the machine and anyway are now no longer used. There has been concern about the possible effects from a VDU on pregnancy in that a number of miscarriages have happened to women working in the same office, but it was shown by statisticians that this is no more than 'a statistical quirk', given that pregnancy and VDU work are common and so is miscarriage, which sadly is the outcome in at least 20 per cent of all pregnancies. Despite this, some health-care workers continue to advise pregnant women to stop working with a VDU during pregnancy. Once this seed of worry has been sown it will be difficult to convince otherwise; so if VDU work is continued, it could be the worry of possible damage to the baby which affects the pregnancy, or bad posture, sitting hunched in a chair all day, affecting placental blood flow. By the time pregnancy is confirmed, if the woman then stops VDU work, the major risk period for foetal abnormality will have passed. Another concern is that the baby will be born deformed or disabled in some way, but again there is no statistical evidence of this. To give peace of mind, temporary alternative work may be arranged, with care to ensure this does not involve such activities as lifting which could hazard the pregnancy. One study has hinted at an increased risk of miscarriage with a full day working on a VDU, but I think any such link could be rather from cramped posture. The increasing use of VDUs will mean for office staff that there is no way to avoid working in the vicinity of VDUs. Then there is the claim that VDU work affects fertility in men as well as women. I know of no evidence of this. Facial rash is a possible health hazard from sitting in front of a VDU. In a dry atmosphere, dust is attracted by the electrical field between the front of the screen and the operator. This dust strikes the operator's face, acts as an irritant and causes dermatitis. This has occurred principally in Scandinavian countries where the air is dry. The remedy is to raise the humidity level. Whether an electrical field could have any effect on health is still a subject for debate and research: there are filters on the market made from wire mesh which earth this out and stop static as well.

For a copy typist or secretary, the word processor with VDU screen is becoming standard office equipment. Advantages over a typewriter include facilities for electronic filing and instant retrieval (mindful of the Data Protection Act and the need to register); quick correction of drafts; and printing of freshly-typed copies of standard letters in seconds. (Managers have been accused of providing sloppy copy [knowing about this ease of retyping], thus adding to the workload.) The machine can be linked to data processing, computing and transmission equipment. Some office work can thus be done very much faster, but some secretaries and typists working in a 'pool', removed from much of the personal contact they had when using the old equipment, complain of boredom and of being turned into robots.

Whilst most word processors have a VDU screen, some manufacturers limit the display to a single line, immediately above the keyboard, of the fifteen words which have just been typed. Other machines allow one letter to be printed whilst the next is being typed. Productivity is greatly increased, but out of all those staff who use a VDU screen, it is likely to be the audiotypist word-processor operator who has most exposure to a VDU screen, given that, even though he or she does not need constantly to look at the screen, it is the natural thing to do and there is nowhere else to look. Even with the ideal workstation, there is the possibility of unfavourable symptoms developing if the working routine is not carefully thought out.

A not-uncommon story is of headaches developing during the afternoon, with concern that the lighting or air-conditioning induces these, but on checking, it is found that natural breaks are not being taken, with the pressure of work increasing to get letters out the same day. Cups of tea and coffee are put on the work station – this is electrically unsafe – and lunch taken early or not at all. As pressures increase, greater concentration is needed, the person blinks less, so the surface of the eye dries and becomes sore; fatigue develops, the neck and arms may ache and so on. All that is needed is to insist on breaks spent, not necessarily in doing nothing, away from the machine. Given the speed of these machines, productivity will not suffer.

PHOTOCOPIERS

These are as much a part of the modern office as the VDU, so it is appropriate to say here there is no hazard to sight from the bright light of a photocopier used without the protective flap down. People may think this light is the same as a welding flash where, if protective goggles are not worn, 'arc-eye' results – a severe conjunctivitis and even temporary blindness from the ultraviolet rays, as in snow blindness. The photocopier produces pure light which causes the pupil of the eye to constrict, thus restricting the amount of light entering the eye, so there is no damage. What is necessary, given that chemicals are used, is that this machine is not tucked away in some airless corner, but is used in a well-ventilated area. (See the section on chemical hazards for possible skin reactions to 'dry' photocopiers.)

TELEPHONES AND AUDIO HEADSETS

These are also part and parcel of office life. A question sometimes asked is whether germs can be transmitted by the earpiece or mouthpiece of a telephone handset or audiotypist's headset earphones; also, can the sounds from the latter affect the ears? The Central Public Health Laboratory attempted to culture germs from many public and office telephones; their conclusion was that the risk of transmitting infection is negligible. In New York, experiments deliberately trying to make the telephone as infectious as possible failed completely. There is no proven case linking a telephone to the spread of an infectious disease. Special cleaning makes a grubby handset pleasanter to use, but from a health point of view, regular routine disinfection of telephones is unnecessary. Usually audiotypists have their own personal headset; the personal stereo type are comfortable, whereas the disadvantage of stethoscope earpieces that go into the ear canals is that they can irritate. Where the earpiece has sponge-covered surrounds, these should be changed regularly when they get dirty. The sound level coming into the headset is adjustable so there should be no very loud noise or any risk of damage to hearing. For the same reason tinnitus – noises in the ears – would not be caused by this work, but existing tinnitus may be more noticeable when working under pressure, as the following enquiry and answers from the British Tinnitus Association explain:

British tinnitus association letter/response

Question: I am employed as a shorthand typist, and it is proposed that I use audio equipment. I have discussed this matter with my doctor, who has supplied a note to my employers stating that the use of audio equipment would exacerbate the condition of tinnitus. I should be grateful if you could help me and let me have your views on this matter and what form such exacerbation would take if I use audio equipment.

Answer: There is no reason at all why the use of audio equipment should exacerbate your tinnitus. The sound that comes out of an audio typing machine can be set accurately with a volume control and in that respect differs in no way from sound in the environment. Indeed with the correct kind of headset it is possible to listen to audio dictation at a sound lower than that experienced in the environment. Many people find that their tinnitus is increased while doing this work but the effect is not a direct one on the ears from the audio machine, but one produced by the stress involved which in turn produces an increase in adrenaline levels in the body. Much tinnitus does respond to stress by a temporary increase in loudness and I am sure that this is the mechanism by which your tinnitus varies.

In summary, there is as yet no definite evidence that operating a Visual Display Unit (VDU) is hazardous to health – but on the other hand it is virtually impossible to prove a negative, that is, to guarantee that a VDU does not cause ill-health. I keep an open mind and a sharp lookout for any mention of any possible health problem in medical journals. Meanwhile, I reassure staff who ring me on this issue.

KEYBOARDS AND REPETITIVE STRAIN INJURY

Keyboards are widely used throughout the world by secretaries, typists, VDU operators and many others, so it was curious that just in Australia there was what amounted to an epidemic of arm pains amongst keyboard users.

An inappropriate title was given to this complaint – Repetitive Strain Injury (RSI) – inappropriate because there is no evidence of injury, and often no sign of strain. Treatment was based on rest; under Australian law, Workmen's Compensation was paid; hence the expression 'golden wrist' (the term 'kangaroo paw' was also used). There is no doubt that arm pains for any number of reasons are common and that keyboard work can highlight these.

Prevention, by reducing an operator's speed of work from 20 000 or more keyboard depressions per hour to 10 000 or less (in theory fewer key depressions reduces the risk of RSI), can frustrate a skilled operator and lead to mistakes.

RSI is not new. In 1830, 'Scrivener's palsy' of the hands was described by Sir Charles Bell in solicitors' clerks; and telegraphist's cramp – wrist pain – affected Morse key operators in 1880. RSI is described in Japan as 'Occupational Cervico Brachial Disorder' and in the USA as 'Cumulative Trauma Disorder'.

A better term is 'Regional Pain Syndrome'. There is aching in one arm which is not related to nerve distribution or to one single anatomical structure. In addition to pain there can be pins and needles, poor grip strength, numbness and diffuse aching in the limb.

The symptoms are worse when the arm is used, either at work or at home, and spreads from the hand up to the neck. But there can be a sharp demarcation between involved and non-involved parts of the arm. There may be sleep disturbance and particular tenderness at specific points.

Other general symptoms experienced by the patient include dizziness, nervousness, anxiety, irritable bowel and tension headaches. Examination shows no evidence of inflammation, wear and tear, nerve or systemic disease and laboratory and X-ray investigations confirm this.

In a way, a person with RSI has the same sort of rheumatic fibrositis pains that are common in the general population, thirty per cent of whom have had arm pain in the preceding twelve months, irrespective of their job. At any one time, ten per cent of the population have pain in the neck or arm.

In practice, RSI tends to affect people who work *at* rather than *with* keyboards, and where the work can be dull, monotonous and boring.

It is important to reassure a person with RSI that the condition is completely reversible and that it is appreciated that the pain of RSI is real and the person is neither malingering nor hysterical. Job satisfaction, which is possible with any job, guards against RSI, as does awareness of muscle tension, the effects of stress and an ability to relax. Doing upper limb and neck exercises may help, as will resting the forearms and wrists on the desk: to have the hands held over the keys for long periods can create muscle tension.

In this country, RSI is bracketed by unions in general and by the General Municipal Boilermarkers and Allied Trades Union in particular, with tenosynovitis, as illustrated in their leaflet 'Tackling Teno'. Tenosynovitis is a localised, painful condition of one or more inflamed, swollen, tender, tendon linings usually at the wrist or base of thumb, where on examination a doctor may feel crepitus (likened to crunching through snow) as the tendon rubs against the inflamed lining. Sometimes a nerve pressure condition known as 'Carpal Tunnel Syndrome' occurs, with initial night-time tingling of the thumb, index, middle and half the ring finger. Also involved can be muscle/tendon or tendon/bone connections on either side of the elbow; pain on the outer side is known as 'tennis elbow', on the inner side as 'golfer's elbow.'

Tenosynovitis and allied conditions are often brought on by unaccustomed repetitive activity. Weekend DIY activities are a notorious cause, for example the twisting action of inserting umpteen screws with a manual screw driver, or painting ceilings, where during the week that person is in a sedentary, clerical job.

Tennis and golfer's elbow can be brought on by activities other than these sports; when associated with tennis or golf they may be caused by an inappropriate grip, or clubs or racquets of the wrong weight. Occupational risks include when first using different equipment at work which needs repetitive movements, and not having proper breaks from such equipment.

In one company where several of the work force had developed tenosynovitis, measurements of blood flow showed a fall with prolonged repetitive activity that was rapidly restored to normal by upper limb exercises. Since these were introduced, there have been no more cases of tenosynovitis.

Rest from the causative activity is part of the management of tenosynovitis and a temporary plaster wrist splint may be provided.

Surgery may be needed to relieve the pressure in Carpal Tunnel Syndrome or in chronic tenosynovitis, but correct and early treatment (mainly ultrasound) will prevent tenosynovitis from becoming chronic. Simple home treatment by intermittent use of an ice pack for five to ten minutes at a time is beneficial. Of equal importance is scrutiny of work practices where these may be implicated in the cause, for example, to make sure the operator sits well supported, at the right height, and is comfortable.

STATIC

Carpeted offices which contain metallic equipment can be plagued by static electricity; the development of voltage between the soles of the shoes and the floor after walking on the carpet is instantly discharged, perhaps with a spark, when a metal object such as a filing-drawer handle or typewriter key is touched. 10 000 volts or more is created in this way and though there are no permanent ill effects, the shock is unpleasant and can also upset sophisticated machinery such as a computer terminal or word-processing machine. Static is a source of ignition in explosive atmospheres.

Carpets (and clothing) made with a high proportion of nylon are most likely to cause 'static.' Many carpets now incorporate an anti-static thread.

With sufficient moisture in the air, build-up of static is unlikely, but in dry conditions, a build-up of static electricity only discharges through metal. Moisture is ensured by incorporating humidification into the air-conditioning system. Where there is insufficient moisture and no humidification from the air-conditioning, the most effective humidification can be provided by a portable electric humidifier which emits a fine spray of water vapour at regular intervals. A bowl of water on the floor, or a shallow tray of water that hangs from a radiator, or even regular sprays of water from a hand-operated pump are simpler and cheaper, but less effective, alternatives. Attractive green plants which need regular watering help maintain good humidity and keep static at bay. An anti-static chemical can be sprayed on the carpet after cleaning, but the effect may not last long. Leather or thick rubber-soled shoes help.

For machinery that is upset by static electricity, another local solution to protect the machine rather than the person is a mains-earthed rubber mat. Some people seem more susceptible to 'static' than others; for them 'anti-static' shoes are available with graphite in the soles.

VENDING MACHINES AND SELF-CATERING FACILITIES

A few companies still employ tea ladies, or rather tea persons, who morning and afternoon visit each office with urn or teapot and trolley. Care must be taken to ensure the person does not use the sort of very large and heavy teapot which gave one tea lady pain from tenosynovitis at the wrist, when her employer apparently failed to heed her complaint of discomfort and her request for a smaller teapot.

Many companies use vending machines which dispense beverages and sometimes snacks as well, at the touch of a button. Often owned by a vending company, day-to-day upkeep and replenishment is, however, the responsibility of the hirer. Regular, efficient cleansing is essential to avoid the accumulation of stale powder, liquid or food in parts of the machine not seen from the outside. A vending machine must not be sited in direct sunlight. When containing cold food, the correct temperature must be maintained and no more than the maximum safe storage time allowed. Vending machines, urns and traditional brew-up methods are potential sites for scalds and for germs. The same preventive principles apply as stated in the chapter on catering hygiene. Regular water samples should be taken from vending machines and checked by public analysts for bacterial count. The inside pipework of the vending machine must be thoroughly cleansed regularly. If this does not happen, there can be a build-up of algae slime in the pipes.

In the absence of vending machines, the area used by employees for the preparation of hot drinks and snacks is often inadequate, possibly a small space in a corner some distance from facilities for washing up. The area must be thoroughly cleaned. The U-bend under a sink contains a reservoir of stale water and is a potential hazard. It should be flushed through with clean water after use and periodically cleansed with an antiseptic.

Particularly in the summer, or if there is no canteen, some staff bring a snack lunch to eat in the office area. Good hygiene in the preparation and clearing-up is needed, plus careful disposal of refuse. Food debris, apple cores, empty bottles and cans should be placed in plastic bags in a pedal bin separate from bins for waste paper to be recycled. The pedal-bin lid excludes flies, wasps and mice.

With work pressures, it is all too easy to let standards slip unless responsibility is given to an individual for regular checking of hygiene standards in the office; this includes the standard of office cleaning; care of facilities in the restroom; defrosting the refrigerator; disposing of out-of-date items, cleaning trays and waste bins, and so on.

A microwave oven as a convenient, quick way for staff to prepare their own hot food is ideal for a staff room, and is not a health risk provided it is looked after in accordance with the manufacturer's instructions. National Westminster Bank's occupational health department produced a poster summarising many of these points.

CHEMICALS – HAZARDS IN USE

Products used in the office environment can harm health: typing-eraser fluid splashing into a typist's eye for instance, or liquid lavatory cleaner into a cleaner's eye. The sniffing of typing-eraser fluid has been reported: mustard is added to discourage this.

EEC legislation – Classification, Labelling and Packaging Regulations 1984 – requires all such chemical products to be marked as hazardous, using standard identification, and covers classification, packaging and labelling of dangerous substances. Similar legislation applies to items that are flammable, explosive, corrosive, oxidising, irritant, or in general harmful to people if swallowed or breathed, or to the environment. Labelling includes the appropriate symbol of hazard.

Chemicals are used everywhere at home and at work. Of over four million listed chemicals, about 100 000 are regularly used at work. Though these are mainly in industrial use, a considerable number are used in the office and in the maintenance of the office, for instance, as cleaning materials. Exposure to some chemicals can harm the body, either immediately, by causing an inflammatory condition such as dermatitis, or much later, usually after long-term exposure, leading to conditions such as allergy or even cancer. Chemicals gain entry to the body by being swallowed, by skin contact or by inhalation. Chemicals get into the air as dust (e.g. asbestos), fumes or gases (e.g. car exhaust), vapours (e.g. trilene) or mists (droplets small enough to stay in the air).

A guideline for the average concentration in the air which is

A little extra care

MICROWAVES
Keep door seals clean –
follow manufacturer's
instructions.

COOKERS
Clean hot plates regularly.
Do not leave unattended
when in use.

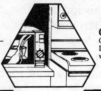

KETTLES
Ensure the kettle isn't
overfull or empty before
switching on. Never fill
when connected to the
electrical socket. Keep
electric lead away
from water. Do not
leave unattended
it may boil dry.

VENDING MACHINES
Clean regularly,
especially the overspill
trays. Ensure that all
internal pipes and
channels are washed
through frequently.

**CATERING
UTENSILS**
Keep crockery,
cutlery and
other utensils
clean; wrap
and throw
away any
which are
damaged.

FRIDGES
Food won't
stay fresh
indefinitely –
discard food
before it
becomes a
health hazard.
Clean and
defrost
regularly.

CUPBOARDS
Overfull cupboards
can be dangerous – keep
neat, clean and tidy –
stack items so that
they will not fall.

TOILETS
Clean up spills promptly
to avoid accidents
and health hazards.
Ensure that all disposable
waste is placed in the
appropriate receptacle
and emptied regularly.

RUBBISH BINS
Use bin liners – wrap
items of food and sharp
objects before disposal.
Never put cigarettes or
matches in them just
after being extinguished.
Empty regularly.

**CLOTHS AND
TEA TOWELS**
Wash and replace
regularly.

**ELECTRICAL
SOCKETS**
Switch off sockets when
not in use, except when
equipment must be kept on
at all times. Tuck leads
away safely where people
will not trip over them.

makes your office a safer, cleaner place to work.

IT'S MOSTLY YOUR RESPONSIBILITY TO YOURSELF AND OTHERS

NWB2169 Nov 87-1

National Westminster Bank PLC, Group Occupational Health Services.
Printed in England by National Westminster Bank, Stationery and Printing Department. 1987 ©

considered harmless on prolonged exposure is produced for some 473 chemicals and revised each year. In the past, American limits were reproduced in this country, with royalties being paid for what is referred to as the threshold limit value (TLV). Now the United Kingdom has its own system, Occupational Exposure Limit (OEL). Gas or vapour is measured as parts per million (ppm); dust as milligrams per cubic metre (mg/m^3), weight count, particles or fibres per millilitre (ml.). OELSs and TLVs are arrived at by experiment, using toxological, scientific evidence, and from experience and for socio-economic reasons. They were initially published by the Health and Safety Executive in Guidance Note EH15; the OELs are now contained in EH40. The average is time-weighted, worked out from fluctuations during an-eight-hour day, forty-year working life. Periodic higher than average levels, 'excursions', must be balanced by lower than average levels, but a laid-down peak must not be exceeded. The abbreviation STEL, 'short-term exposure limit', denotes where, within the OEL, exposure to a higher concentration as shown is considered safe for up to fifteen minutes. For some chemicals OEL must not be exceeded; this is denoted by the prefix C for ceiling.

As an example, a possible hazard from 'dry' photocopiers is dermatitis from the chemicals used which, like other solvents, have weak skin-defatting properties. Prolonged skin contact will therefore cause a primary dermatitis. Also possible is an allergic skin reaction, which could be confirmed by 'patch-testing' by a dermatologist.

ASBESTOS HAZARDS

Asbestos is only hazardous to health if it is loose in the air as dust and is breathed into the lungs or swallowed. Once asbestos is in the lungs it cannot be removed. An asbestos-related illness takes many years to develop. Asbestos insulates, is fire-resistant, and has been used in the building and ship building industries for many years. Uniquely asbestos, which is mined from rocks in Russia, Canada and South Africa, is formed by silicate crystals which recrystallise into long, flexible, fibre shapes which can be spun. Three types of asbestos have been mainly used in the UK – white chrysotile, brown amosite and blue crocidolite – so called because of their tints when mined. All three types can be hazardous to health.

The cause of asbestos-related illness is microscopic slivers of asbestos fibre which flake off loose asbestos and are inhaled. The body's reaction coats them with a golden sheath, 'asbestos bodies', seen on microscopic examination of sputum. The presence of these does not necessarily imply lung disease. The size of the fibre is critical – some penetrate deep into the lung. The blue asbestos fibre is especially fine – a straight, needle-like crystal. Undisturbed, asbestos does no harm. It is mainly people who have worked, unprotected, closely with asbestos who are at risk. Such a person who also smokes is at particular risk, and is as much as fifty-five times more likely to develop cancer of the lung compared with a nonsmoker who has not worked with asbestos.

There is a great deal of publicity and misinformation about asbestos. It is most unlikely that an office worker in a modern building will have been exposed to the risk of inhaling asbestos dust, but special care should be taken by maintenance staff, plumbers, electricians, and others engaged in refurbishing, where electricians may drill partitions containing asbestos, or, for example, hot-water-pipe insulation is to be renewed and the old lagging was asbestos fibre. High temperatures can make the fibre flake, and stripping lagging creates asbestos dust. Also requiring great attention is where, particularly in such buildings as stores and warehouses, ceilings have been sprayed with an asbestos-cement mixture as insulation against the spread of fire. Where this mixture has not been sealed, and given certain conditions, in time flakes of asbestos could fall from the ceiling and blend with existing dust. Sealing will prevent this.

Asbestosis, fibrosis or scarring of the lungs with restriction of breathing, causes breathlessness on exercise with abnormalities on clinical examination, on chest X-ray and on testing lung function.

Exposure to either very small amounts of dust, or for a very short time, will not cause asbestosis, which only develops in a person who has been working closely with asbestos, inadequately protected, for years. A complication of asbestosis is cancer of the lung or throat, or of the lining of the chest wall or abdominal cavity (mesothelioma). As mentioned, the incidence of lung cancer is much higher in asbestos workers who smoke. Mesothelioma can also occur without asbestosis, many years, usually twenty or more, after exposure to asbestos dust. Even then, exposure has been several weeks or months to large amounts of asbestos dust.

Asbestosis and mesothelioma in those who have been working with asbestos are 'Prescribed Diseases', assessed by Regional Pneumoconiosis Medical Panels which can award between ten and one hundred per cent disablement benefit and, if there are dependent relatives, a death benefit when death occurs.

A successful claim may be made under common law by someone who has been working with asbestos, if it can be proved that the employer was negligent in maintaining the standards required by law. Such a claim must be started within three years of the claimant being aware of the diagnosis. Beyond this time, the claim would be 'statute-barred'. What can be difficult is proving the diagnosis of asbestosis during life. After death, comprehensive analysis of the lungs is possible at *post mortem* and diagnosis therefore more certain.

Wherever asbestos is discovered to be unsafe and has to be removed, very special regulations and great precautions apply, and a specialist firm is brought in to do this work.

Under legislation, the *Asbestos Regulations 1969*, there are very special requirements before any work on asbestos is started. Notice must be given, the insulation must be soaked with water to reduce dust, and protective equipment must be used, including approved dust respirators and protective clothing. There is an HSE approved code of practice, 'Work with Asbestos Insulation and Asbestos Coating', to protect workers dismantling asbestos lagging or sprayed-on asbestos, and anyone in the environment where this work is being carried out. The work area is sealed off and notices warn that only those using breathing apparatus – negative-pressure dust extraction – and wearing protective clothing may enter the area.

From August 1984 only licenced contractors may remove dangerous asbestos material. Any removed asbestos must be put in heavy-duty polythene bags marked 'asbestos – do not inhale dust'.

Workers must remove all protective clothing before leaving the site. Showers are provided between the work area and where personal clothing is left, to ensure that no dust leaves the site. No protective clothing is taken home; formerly the wife who washed her husband's asbestos-dust laden clothes for years was at risk of developing mesothelioma years later.

Other insulation materials include man-made mineral fibres (MMMF). These have been used for over forty years. They can cause transitory skin irritation; upper-respiratory-tract irritation has also been described, but there is to date no clear evidence of any lung disorder resulting from exposure to MMMF.

CONTROL OF SUBSTANCES HAZARDOUS TO HEALTH

Nineteen new regulations, ten schedules, and a series of codes of practice, with the collective title 'Control of Substances Hazardous to Health', known for short as COSHH, were first outlined by the Health and Safety Commission (HSC) in a consultative document in August 1984. The purpose of COSHH is to safeguard the health of anyone exposed to 'substances hazardous to health arising out of, or in connection with, work activities'. There was a huge response to the request for comment: a working party and task groups produced a new draft, which, having been approved and passed by Parliament, brings COSHH into force in 1989.

These regulations cover all substances classed in the *Packaging and Labelling of Dangerous Substances Regulations* as being corrosive, irritant, toxic, very toxic, or in any other way harmful. Also included are 'substantial' amounts of dust in the air; human pathogens (germs harmful to human beings), and a round-up of any other substance arising from work which may be hazardous to health.

Existing legislation had been too complicated and too general, and was assumed to relate to factory work alone, whereas ill health may be caused by occupational exposure to toxic substances in any work setting – the office as well as the factory floor.

COSHH provides one set of regulations; to establish principles and standards; to take account of known risks of exposure; to provide flexibility, to take account of future use of any new toxic substance, and to ensure compliance with the law.

COSHH is aimed principally at the employer's duty to his or her employees, and also to anyone else generally affected by the work activities, either immediately or after a time lag. As with the whole of HASWA, employees are expected to co-operate, to report defects, and so on.

To fulfil what COSHH requires, involves three steps – assessment, control and monitoring. Assessment is the crucial first step – get that wrong and the entire strategy collapses. Assessment can mean anything from merely looking at information such as the manufacturer's data hazard sheets, to a full environmental survey by an occupational hygienist.

Control should be from the source rather than by use of personal protective equipment (PPE) – a golden rule in occupational medicine but tempered, as in all of HASWA, by the phrase 'so far as is reasonably practicable'. Where possible, the appropriate principle of elimination, enclosure, or exhaust ventilation should first be applied, and where PPE is needed, it must be properly maintained and the employees taught how to use it.

If monitoring is shown to be needed, records must be kept – and for a very long time: fifty years, if linking exposure to effect. Records must also be made available to the employee, who must also be made fully aware of the hazard and what the employer is doing about it.

CONTAMINATED PREMISES AND MAIL

There is a potential risk of bacterial infection to staff who reoccupy flooded premises, particularly where there has been leakage of sewage. Waterproof protected clothing and gloves should be worn, and no eating or drinking on the premises is the rule. At the end of the working day, and before eating or drinking, hands should be thoroughly cleaned and finger nails scrubbed.

There should be no health risk to staff from the unpleasant experience of dealing with mail which has been lying in the front-door letterbox over the weekend, in a pool of urine from vandals. Wear gloves when opening and drying the letters. Raise the level of the letterbox to make a repeat of such an act impossible.

12 Catering

FOOD POISONING

Staff restaurants providing meals at work – 'in-house' or using outside caterers – are a welcome amenity. The company must do all it can to ensure the food is wholesome.

Food poisoning can be caused by eating poisonous plants, fish or chemicals, but the commonest cause is contamination of food by certain bacteria or the poisons they produce in food. Food-poisoning germs need nourishment, warmth and moisture to grow, meat and poultry dishes are the main sources.

Annual statistics show that 80 to 90 per cent of reported incidents originate from cold or reheated meat and poultry dishes, including soups, stews and pies. Milk and cream products such as custard, trifle and creamery products also feature; fish is rarely implicated unless eaten cold after long storage without refrigeration, or mixed with other ingredients, as in fish cakes and pies.

Food poisoning is deceptive as even food teeming with germs usually looks, smells and tastes alright. Spoiled, mouldy, decomposing food is easily detected and avoided as its smells and tastes 'off'. Likewise, foreign objects in food are usually readily seen – this is the reason for catering staff's adhesive dressings being blue.

Food poisoning is a notifiable illness. Where a company doctor or nurse is employed, it is likely that anyone with food poisoning resulting from a meal in the staff restaurant will first present themselves at the Health Services Unit. The doctor or nurse will then liaise with both the management and the Environmental Health Officer of the local authority.

The incidence of reported food poisoning is rising (over 13 000 cases a year). Incidence is partly dependent on the weather – it peaks in heat waves and declines in winter. For every notified case, there are likely to be several that are not reported, for such reasons as misdiagnosis as a virus infection symptoms so mild or transient that a doctor is not consulted.

For food poisoning to occur there must be a sequence of:

(1) The presence of germs, and food in which these germs thrive.
(2) Transfer of germs to the food via contaminated equipment, other food or fingers.

(3) Conditions favourable for the multiplication of these germs.
(4) A susceptible person who then eats the food.

Food poisoning is avoided by cleanliness: clean food, handled by clean people, using clean equipment in clean premises. People with sloppy hygiene who excrete food-poisoning germs in their stools, do not wash their hands after using the toilet, and then handle food, may transfer the germs to the food, where they multiply. The food-handler can break the food-poisoning chain by carefully observing simple rules about kitchen and personal hygiene in the preparation and storage of food.

Salmonella, listeria, shigella, Escherichia coli, staphylococcus, clostridium and *Bacillus cereus* are some of the bacteria responsible for food poisoning (see Table 3). There are more than 1700 types of *salmonella* which can cause illness, including gastroenteritis, from food poisoning. Up to forty-eight hours after the meal there is sudden malaise, headache, sometimes a fever, then nausea, vomiting, profuse diarrhoea and abdominal pain. A mild case settles down in two days, but often the bowels are not back to normal for up to ten days. Stool cultures confirm the diagnosis, plus, if possible, finding salmonella in the left-over food and raw materials, provided these have not been thrown away, which so often happens. The GP must be told that the patient works in a kitchen.

Listeria causes similar general symptoms to salmonella without the bowel disturbance.

Clostridium perfringens is frequently present in the gut of humans and animals, and produces heat-resistant spores which survive boiling. Poultry and other meats are often contaminated with the animal's faecal matter which contains these spores, and whilst pressure-cooking kills the spores, conventional cooking activates them, so they germinate and then multiply rapidly if the food is left at warm temperatures. The swallowed germs produce a toxin causing abdominal pain and eight to twelve hours later profuse diarrhoea, which lasts one to two days. The illness is avoided by rapid cooling and cold storage of meat and other foods not eaten immediately after cooking. This prevents growth of the germ.

Clostridium botulinum causes botulism, fortunately very rare in the United Kingdom. The germ produces a lethal toxin in food under strictly anaerobic (absence-of-air) conditions. The food is usually imperfectly preserved. The toxin affects the nervous system and causes paralysis and death in many cases.

Table 3 Summary of common causes of food poisoning

Germ	Likely source of infection	Illness caused by	Incubation	Main symptoms
Salmonella	Poultry, raw eggs, precooked meat eaten cold	Bowel invasion	12 – 48 hours	Fever Vomiting Diarrhoea
Clostridium perfringens	Reheated meat	Toxin in gut	8 – 12 hours	Abdominal pain Diarrhoea
Shigella	Person-to-person spread, directly or via food	Bowel invasion	48 – 96 hours	Fever Diarrhoea
Eschericia coli	Many foods eaten abroad	Toxin in gut	24 – 48 hours	Diarrhoea Abdominal pain
Staphylococcus	Handled cold food	Toxin in food	2 – 8 hours	Vomiting Collapse
Bacillus cereus	Reheated rice	Toxin in food	1 – 5 hours	Vomiting
Campylobacter	Poultry	Toxin	12 – 48 hours	Fever Abdominal pain Diarrhoea Collapse
Listeria	Poultry, soft cheeses (unpasteurised), pre-packed salads	General effects	48 hours	Flu-like illness: headache, fever, muscle pains; risk groups: the elderly, babies, toddlers, pregnant

Shigella can be food-borne and causes a prolonged diarrhoeal illness after two to four days' incubation.

Certain types of *Escherichia coli* cause diarrhoea in adults as well as infants. Incubation is about twenty-four hours, then there are griping stomach pains with diarrhoea and nausea, but rarely vomiting. This germ is responsible for many cases of traveller's diarrhoea.

Certain strains of *staphylococcus* found on the skin multiply in moist food and produce a toxin. *Staphylococci* thrive in custards, cream and cold meats. After two to eight hours there is sudden vomiting, diarrhoea and even collapse if a lot of toxin has been swallowed. Recovery begins about six hours later and is complete within two days. The source is likely to be a member of the kitchen staff with a septic skin disorder, discharging nose or ear, or something similar.

Bacillus cereus produces a toxin in cooked food, particularly in cereals like rice which are then left to cool or kept warm for many hours. Rapid onset of profuse vomiting results.

Giardia lamblia (a parasite) swallowed with contaminated water causes chronic diarrhoea and weight loss. *Amoebic dysentery* may be contracted in the tropics. Food-poisoning germs grow like wildfire, dividing every twenty to thirty minutes. Apart from causing acute illness, they can linger in the apparently cured, who become symptomless carriers and, if food-handlers, can still cause food poisoning in others.

The main sources of these germs are:

(1) Raw food: germs (salmonellae) cross from raw to cooked foods on hands during food preparation, or via surfaces, utensils, machinery, cloths or other kitchen equipment. Even in the most perfect kitchen, catering staff are reluctant to give up the multi-purpose ubiquitous cloth which hangs from the waist.

(2) Spores in the raw materials may survive cooking, germinate and multiply in slowly cooling, unrefrigerated food left out for hours in a warm kitchen.

(3) The nose, the skin, especially a septic area harbouring staphylococci are potential danger spots if touched because the hands may then transmit germs to cooked food eaten cold. Hands are anyway difficult to free from staphylococci even by scrubbing. Food-handlers with skin sepsis should not touch cooked foods; indeed they should stay out of the kitchen and servery.

(4) An impure water supply is a serious hazard, not only to those drinking it but also because water is used in the kitchen.

(5) Flies, cockroaches and other insects, rodents, cats and dogs carry germs from faecal matter to food and from food to food.
(6) Dust which contains spores may contaminate food.
(7) The food-handler who does not regularly and frequently wash his or her own hands before and after handling foodstuffs and after using the lavatory.

FOOD STORAGE AND PREPARATION

The storage and preparation of raw food must be completely separated from that of cooked food, as must be the equipment and utensils used for each.

Foods not eaten freshly-cooked must be refrigerated as soon as possible. The times between cooking and cold storage, and cold storage and eating, should be as short as possible. Reheated food from a refrigerator should be boiled or otherwise thoroughly cooked. Great care is needed with surplus food.

The cook/chill method of food preparation (cooking followed by rapid chill, limited storage and thorough reheating) is gaining favour as being highly cost-effective from the point of view of both manpower and time as well as materials, because of low wastage. It is essential that specially-trained staff are employed for this work, as food poisoning is a danger if the steps outlined are not meticulously followed. Temperature probes are also necessary, to confirm the correct low and high temperatures have been reached.

In a working kitchen it may be impossible to keep everything spotless all the time, but at least accident risk can be reduced by careful disposal of breakages and prompt attention to spillage of both liquid, which causes slips and falls, and food which attracts vermin. Rodent bait, or tubes in the runs coated with poison which gets on to the animal's fur and is licked off, are suitable. Special ultraviolet light fittings attract and kill flies. Regular in-depth cleaning is needed as well as daily routine.

The main meal provided is usually lunch. Large companies may provide restaurants, have sophisticated kitchens with skilled staff, and employ catering managers. Small companies may provide a kitchenette in the rest room with such self-catering facilities as a small electric cooker or microwave oven.

It is important to follow the manufacturer's instructions about how to install, use, clean and maintain the oven. For a microwave oven it is vital to keep the door seals in good repair, and prudent to arrange for an annual radiation leak check by a qualified person. Certain items like metal or foil dishes will ruin a microwave, and heating as it does from the inside outwards, there is a risk of burning or scalding the mouth with a cooked item that appears outwardly to be deceptively cool.

Also provided will be an electric kettle (use a flexible, that is, coiled lead and a sleeved plug), and a refrigerator. An icebox in the fridge is useful to keep icepacks in for use in first aid; but in that injuries should be rare, it is perhaps best for the fridge to be self-defrosting to overcome the common failure to defrost it regularly. A hygiene poster in the rest room acts as a reminder of good practices.

In a small company, senior staff may entertain in-house and use either outside caterers or have a small kitchen for preparation of cooked meals. In this latter instance, the manager is responsible for ensuring good hygiene standards for the preparation and serving of cooked meals, and the following section suggests what to look out for when there is the suspicion of poor hygiene or possible food poisoning. (One way to exonerate a catering unit being blamed for causing illness in staff who have eaten there is to keep portions of every item served, chilled for forty-eight hours. Should more than, say, five customers suffer apparent food-poisoning, the relevant portions are analysed by a laboratory and the Environmental Health Officer advised of the results.)

FOOD PREPARATION AREAS

What follows covers hygiene in the kitchens of 'in-house' catering units, preparing meals for the employees of large companies. Many of the principles also apply to staff rest rooms provided with fridge and cooker, where staff prepare and eat their own meals.

Professional caterers should be well aware of the rules, but other staff may not be, and hygiene may be no one's particular responsibility. Hygiene can come low in the tasks priority list compared with running the business.

All food-handlers, kitchen cleaners and porters must know, and be regularly reminded by way of staff meetings and posters, why good hygiene is essential. Hands should be washed not only as part of toilet hygiene but also between tasks, and particularly after touching raw meats. Catering staff should be encouraged to study for and take the relevant food hygiene examinations. It is compulsory to display 'now wash your hands' notices in catering unit lavatories. Lavatory bowls should have lids and ideally foot-operated flushes, as a handle or chain can be contaminated. Wash-hand basins must be provided in the kitchen as well as the toilet areas, with running hot water, liquid soap from a dispenser rather than a bar under which bacteria can survive, and a plastic nailbrush (nylon bristles). Nailbrushes should be disinfected regularly by boiling or disinfectant and stored up-ended.

Towels should be paper disposables or 'continuous' pull-on rollers. Water flow in wash-hand basins should ideally be controlled by pedal or elbow taps, as conventional taps become contaminated. Antiseptic-barrier handcream from a wall-mounted dispenser lessens the risk of dermatitis, preventing skin roughness and cracks which harbour germs.

Food-handlers should wear light-coloured protective clothing, provided in sufficient quantity, and backed by adequate laundry facilities to allow frequent changes.

Despite these measures, staphylococcal germs remain on the hands; so foods that support the growth of staphylococci such as cold cooked meats, cream and custard confectionery, must not be touched.

Dishcloths should be paper disposables; a general-purpose mutton cloth transfers germs and becomes heavily contaminated. If these or drying-up cloths are used, they should be washed out and boiled daily.

Modern surfacing materials, composite flooring, plastic or laminated table tops, hard rubber boards, stainless steel and vitreous enamel all help to make it easier to keep a kitchen clean and hygienic.

Use of tiles facilitates cleaning, but cracked tiles on floors or walls harbour germs and must be replaced. Cracks in walls, floors, table-tops, and dirt-traps around pipes, plus areas behind equipment, if that equipment cannot be moved out on castors, must be sealed off. The junction between worktops and walls, and between walls and floors and ceilings, should be coved to facilitate cleaning.

Fan air-extraction units are dust and dirt traps and require regular, thorough cleaning. Failure to do this is not only a hygiene risk, but also causes sub-standard working conditions, as the area becomes hot and smelly. Ultraviolet light behind an electrified grid is used to attract and kill flies and other insects. Great care must be taken when using an insect or fly repellent to avoid contamination of food.

Implements must be scrupulously clean. Wooden surfaces, including shelving and the traditional chopping board, are difficult to clean and disinfect and should be avoided.

Floors should have a non-slip surface, and be scrubbed as often as necessary to maintain cleanliness, then dried thoroughly and kept dry, with spillages cleaned up immediately. Drains must be cleaned regularly and particular attention paid to scraps and grease in gratings, traps and bends. Flies breed in dirty drains and food scraps attract rodents. Great care must be taken in the use of disinfectants to avoid contamination of food: chlorine compounds are effective and safe.

Walls, doors, windows and light fittings must be regularly washed – (the frequency determined by regular inspection) to remove grease and dust. Heavy static equipment, such as ranges and steamers, need daily cleaning, with programme cleaning of gas rings and jets, grease traps, canopies, and so on. Cooking equipment must be thoroughly cleaned before and after use, washed in hot water containing soda or detergent, disinfected in boiling water or hot water containing disinfectant, then thoroughly dried and stored in a current of air. Stores must be scrupulously clean and fly- and rodent-proof.

Foodstuffs should not be packed so tightly in a refrigerator as to prevent circulation of cold air. Hot food must be rapidly cooled to room temperature and then put in the refrigerator (if put in hot, it raises the temperature and spoils food already there). Foods and liquids should be placed in moisture-proof wrappings or in covered containers that protect from dehydration or contamination. Refrigerators, on castors for ease of moving to clean behind them, must be regularly defrosted and cleaned.

In the swill area there should be refuse bins with lids that fit. The area should be rodent-proof, and have a concrete or similar type of floor with good drainage for easy cleaning. The floor must be scrubbed, food scraps removed and bins emptied each day. The bins are then washed out and disinfected. Steam injection is the best sanitising agent.

In the dining room the same attention to detail is required as for the kitchen. Waiters and waitresses must wear appropriate protective clothing, that is, a uniform. Floors and chairs should be cleaned each day, and a proper system used for clearing and cleaning tables. All cutlery, crockery and glassware must be heat-disinfected after washing. When a washing up machine is available, a disinfection sequence should be built into the programme.

KITCHEN AND SAFETY NOTES

Scalds and burns

Immerse at once in cold running water for 10 minutes. Cover with a clean dry dressing. Refer to skilled aid. Use eye wash bottle for chemical in the eye.

Falls

Do not overreach. Never run in work areas. Clear up spillages. Report unsafe conditions. Wear sensible, safe shoes.

Cuts

Wash your own hands, cover any cut or graze on your hands, wear disposable gloves. Apply pressure to stop bleeding. Wash and dry minor cuts. Apply blue adhesive waterproof dressing. Remember your knife drill.

Electric shock

Switch off. Resuscitate. Treat burns. Remember electricity and water are lethal.

Hands

Wash before and after handling food and after using the toilet. Use nail-brush for nails only.
Hotter water = cleaner pans. Change cool or dirty water. Wear gloves and protect eyes when using corrosive agents.

Skin

Do not spread infection. Report all skin, throat and bowel disorders at once. Keep hair covered.

Don't give germs a chance

Germs breed in warm food. Serve cooked food hot, or cool and then refrigerate. Separate raw or cooked foods and the utensils used. Keep everything clean. Disposable cloths are best. Do not smoke. Never cough or sneeze over food.

Rubbish

Flies are a danger to health. Keep dustbins covered. Dispose of all waste properly and frequently.

Lifting

Repeated small stresses to the back, such as over-stretching, damage the back as surely as one major injury; both may cause permanent damage:

- At all times do not stoop – kneel rather than bend.
- Think before you lift.
- Don't attempt too much on your own.
- Face the direction in which the load will be moved.
- Stand with feet apart one foot slightly in front of the other for good balance.
- Lift, bending hips and knees with a straight, not necessarily vertical back. Take the same care putting the load down.
- At all times keep the load as near to you as possible.
- Lift items singly or ensure a balanced load. When lifting a heavy bag, place one hand under the bag and the other hand grasping its neck.
- If possible, divide a heavy or bulky load or use a mechanical aid.

A SUMMARY OF INSTRUCTIONS – be as positive as possible

Dos

DO keep raw and cooked meats separate, and also the implements, dishes and cloths used for each.

DO defrost frozen food adequately, especially meat and poultry. Without this, cooking is incubation rather than sterilisation.

DO cool quickly all cooked food not to be eaten immediately.

DO take extra special care in warm weather.

DO ensure that waste is quickly and correctly disposed of.

DO keep working surfaces and floors clean and dry – wipe up spills promptly.

DO   wear special clothing and safety footwear when provided, and hats which enclose the hair rather than sit on it.

DO   wash your hands regularly and
(1) Before handling food, particularly cooked food.
(2) When the hands are dirty.
(3) After:
    (a) using the lavatory;
    (b) handling meat and poultry;
    (c) disposing of refuse;
    (d) nose blowing.

DO   use paper tissues.

DO   take extra care when dealing with hot liquids or hot stoves.

DO   know where to find the first aid box and the first-aider or appointed person in first aid.

DO   cover any cut, burn or sore, however small, with a blue waterproof dressing, and report any injury, skin infection or rash to the manager.

DO   make sure you know what to do in the event of fire: where to find the right firefighting equipment, e.g. chemical extinguisher, fibre-glass smother blanket, and know how to use them.

DO   wear rubber gloves with clean, dry, cotton liners plus protective goggles when using chemicals like oven cleaners.

Don'ts

DON'T use unguarded machinery.

DON'T clean electrical equipment still connected to the mains.

DON'T touch cooked foods.

DON'T smoke in a food room (illegal), lick your fingers or bite your nails.

DON'T leave lids off waste bins.

DON'T lift a heavy pan, urn or tray on your own – get help.

Part III

Sickness and Accidents

13 Sickness Absence

PATTERNS OF INCIDENCE

Over the past twenty-five years, the pattern of causes, duration and frequency of sickness absence, that is absence from work attributed to ill health, has changed. More working days are lost through ill health than through industrial disputes. Sickness absence rates in the United Kingdom average in the order of 16.5 days per employee per year. This is for GP-certified absence, that is, more than seven days. No national statistics are available for self-certificated, short-term sickness absence (STSA) of up to seven days.

The number of days of certified incapacity for men increased by thirty per cent in the twenty years up to 1975. More than 25 per cent of the total occurred amongst those aged sixty to sixty-four, forty-seven days' absence a year. By contrast, the highest proportion of female sickness absence was amongst the age group twenty to twenty-four.

In Britain there are considerable regional differences in the levels of sickness absence from work, and as the historic table below shows, levels and also differences increased for most regions between 1953/4,

Table 4   Sickness absence by region of Britain: days of certified incapacity per man (rates calculated on the basis of the average male population at risk)

	1953–4	1971–2	1974–5
Wales	20.2	30.6	31.8
Northern	16.4	24.0	25.3
Yorkshire and Humberside	14.1	20.4	22.0
North-West	14.1	19.8	21.9
Scotland	14.7	19.4	19.8
East Midlands	11.9	15.0	16.8
South-West	12.5	15.3	17.0
West Midlands	11.9	13.8	14.3
East Anglia	10.1	11.6	10.0
South East	10.1	10.1	9.7
Great Britain	12.8	15.9	16.5

Source: DHSS

Table 5   Certified sickness absence: working days per person in 1972 for England and Wales

Socio-economic group	Males	Females
I	3.0	3.4
II	6.8	4.9
III (non-manual)	5.3	6.5
III (manual)	10.0	8.3
IV	12.8	7.4
V	22.6	8.5
All	9.3	7.0

Source: *General Household Survey*, Office of Population Censuses and Surveys.

1971/2 and 1974/5. South-East England and East Anglia went against the trend of increase, while Wales and Northern England continued to head the league table.

Sickness absence rates also vary according to sex and socio-economic group (see Table 5). There have been significant changes in reasons for consulting GP's with a considerable increase in consultations for *nervous disorders*, particularly *depression*, for which half a million people are treated each year; as many as a third of those consulting also complained of *anxiety*. 26 per cent of consultations were for *respiratory disorders*, with an increase for *bronchitis* (acute, 6 per cent, chronic, 1 per cent), and *colds* and *throat infections* (7.5 per cent); *asthma* accounts for 1 per cent; disorders of the *circulatory system* showed an increase for hypertension and heart and brain blood-vessel disorders, such as *heart attack* and *stroke*; *skin disorders* total 11 per cent with an increase for *eczema*; *bone and joint disorders* total 9 per cent and *migraine*, 7 per cent. *Rheumatoid arthritis* affected 500 000, *epilepsy* 300 000; *multiple sclerosis* 50 000, and *Parkinson's disease* 70 000.

A striking feature has been the increase in absences for relatively minor symptoms or non-specific, poorly-defined conditions (such as debility, headache, backache), where diagnosis is largely dependent on what the person says. It seems that some people now regard minor symptoms as reason enough to stay away from work.

Sickness absence statistics, from a I per cent sample of claimants, are somewhat misleading for several reasons:

Calculations are based on a six-day week, regardless of Bank

Holidays although most people are contracted to work less than 250 days each year.

A third of the total applies to those unfortunate long-term sick who are virtually unemployable.

Short spells of self-certificated short-term sickness absence (STSA), that is, for less, than four days, 'waiting days', may not have been reported.

Sickness absence in the armed forces is excluded, and of mariners at sea, most non industrial civil servants and Post office employees.

Many married working women and some widows are uninsured by choice for sickness and invalidity benefit, and so their absence for other than industrial accidents is excluded.

The Statutory Sick Pay (SSP) book gives a useful list of doctors' shorthand abbreviations – initials which might be used as diagnoses on medical statements, and which would leave the employer none the wiser as to how long that person might be away, or when an absence was lasting an unexpectedly long time.

CHD	–	Coronary heart disease
CVA	–	Cerebrovascular accident
D&C	–	Dilatation and curretage (of the cervix)
DS	–	Disseminated (multiple) sclerosis
DU	–	Duodenal ulcer
D&V	–	Diarrhoea and vomiting
FB	–	Foreign body
GU	–	Gastric ulcer
IDK(J)	–	Internal derangement of the knee-joint-(torn cartilage)
IHD	–	Ischaemic heart disease
LIH	–	Left inguinal (groin) hernia
MI	–	Myocardial infarction (heart attack)
MS	–	Multiple (dissiminated) sclerosis
NAD	–	No abnormality detected
NYD	–	Not yet diagnosed
OA	–	Osteoarthritis
PID	–	Prolapsed intervertebral disc (in the spine)
PUO	–	Pyrexia (fever) of unknown origin
RIH	–	Right inguinal (groin) hernia
URTI	–	Upper respiratory-tract infection
UTI	–	Urinary tract infection
VVs	–	Varicose veins

This list does not of course include every abbreviation used, and others I would add are DJD, for degenerative joint disease and synonymous with OA, RA for rheumatoid arthritis, and HH, for hiatus hernia.

Terms used amongst doctors and unlikely to be seen in print, belonging more to such books as *Doctor in the House*, include SRI – seems rather ill, GOK – God only knows!, and WOOS – work out own salvation!

Somewhat meaningless in terms of prognosis are terms like 'malaise' and 'nervous debility'. 'Coryza' is the medical term for the common cold; if everyone stayed off work with a cold, offices would be de-populated!

This book is not intended to give details of every common illness, but information will be given about a number of the conditions mentioned earlier and also some of the other conditions which are frequently referred to company doctors, with especial reference to the all-important prognosis.

Managers may grumble about high STSA rates, but despite the increase referred to earlier, it is a wonder that rates are not higher. A number of surveys in representative communities show that, in the previous two weeks, most of the people questioned had felt sufficiently unwell to seek some remedy, either from their GPs or direct from chemists.

A survey of interviews with adults of working age in South London showed that, whilst 95 per cent mentioned a health complaint in the previous two weeks, and 20 per cent consulted a doctor, only 8 per cent had taken time off work. A *General Household Survey* by the Office of Population Censuses and Surveys, published by HMSO and based on 250 000 interviews, confirms this. Eight out of ten men and women claimed to suffer from ill health; 43 per cent of men and 57 per cent of women had taken medicine of some kind in the previous two weeks, and a third of men and 20 per cent of women were constantly taking some form of treatment. As might be expected, the older the age group, the more the complaint of chronic ill health, but even in the sixteen to forty-four age group, almost 50 per cent of men and 60 per cent of women claimed a recurrent health problem. Only 23 per cent of men and 15 per cent of women considered themselves to be completely healthy.

A company doctor, though not directly employed to reduce sickness absence, and in no way having any disciplinary role, may, if given the opportunity and computer facilities to analyse sickness absence records, identify situations for further investigation.

Analysis of sickness absence on individual, office, department, building, area, region, and national levels may reveal patterns of prevalence for particular conditions, which can be compared with peer groups elsewhere. For example, the value of a health screening programme can be judged or an occupation-related illness such as humidifier fever may be identified. This is no substitute for visiting, which is the only way symptoms not bad enough to cause absence can be identified.

Health education and other preventive medicine measures, such as the availability of lifestyle health screening, may help staff stay healthy. In some companies, health education takes the form of talks or films for staff during the lunch break, on such topics as back pain, heart attacks, foot care, obesity and dental care. Other companies, with a more scattered work force, cover these subjects in personnel bulletins or an 'in-house' newspaper. An initiative started by the late Health Education Council, taken up by many companies and embracing many of these ideas, is the 'Look After Yourself' campaign, and additionally there is the Health Education Authority/DoH 'Look After your Heart' campaign.

Statistics show that sickness absence varies with the sex, age and status of the employee. That it is highest amongst young married women, possibly reflects split loyalties between employer, home and family, and least in the male manager, probably linked to career aspirations and job satisfaction. Sickness absence among married women is almost double that of men. This is not necessarily because married women, or their families, have more illness; statistically the higher the grade the lower the sickness absence rate, so the rate may relate to the generally lower occupational status of women. Against this, however, is the observation that single women, divorcees and widows have sickness absence rates comparable with those of men.

Research by the late Dr Peter Taylor, comparing sickness absence between European countries, showed that the UK rate at the time (fifteen days per worker) was the same as in West Germany and Poland, and was exceeded in Czechoslovakia (sixteen days), Sweden (eighteen days) and Holland (twenty-one days). In Holland, 400 special doctors with no clinical responsibility become involved after two weeks' sickness absence. These doctors, rather than the doctor giving treatment, then determine sickness absence claims. This is an expensive system and curiously sickness absence rates in Holland are amongst the highest in Europe.

In Sweden, the first week of sickness absence is covered by formal self-declaration plus surveillance by Sick Fund Officials. (After one week GPs certify inability to work.) The number of short spells of sickness absence (under a week) in Sweden were double that in Britain when UK self-certification was for three days only. Some years ago, during a period of dispute in the DHSS, many UK GPs stopped issuing DHSS certificates of incapacity for work. During this time there was a 'trust-the-patient' scheme, in which the patients self-certified periods of sickness absence (which did not involve signing their own DHSS certificates), and sickness absence increased.

In Communist countries, factory doctors certify sickness absence of ambulant workers. Sickness absence rates are similar to this country. In Russia, 'prophylactoria' treatment units are sited at the work place, where those with not totally incapacitating conditions can be treated after work, in the evening and at night; sickness absence has significantly declined.

In the Soviet Union a general practitioner may not issue a sickness certificate for more than three days, except in an epidemic of influenza, and cannot authorise sick leave for more than six days at a time for any one episode of an illness. For an extension the doctor must have the consent of a Committee – a medical consultative commission convened as necessary at the clinic or hospital. The commission may also alter conditions of work and arrange transfers. Another commission deals with invalidity.

That sickness absence in the Soviet Union fell from 15.2 days in 1965 to 13.5 days in 1974, whilst the level rose in other European countries, reflects the tight control of certification and the great importance attached to minimising sickness absence. What is not known is whether premature return to work occurs and if so whether harm results.

SICKNESS ABSENCE PROCEDURES

Since 1983, for most people in employment who pay Class One National Insurance contributions, the system of State Sickness Benefit for up to twenty-eight weeks has, in two stages, been replaced by SSP from the employer, who recovers the gross sum by withholding this from the National Insurance contributions. After an interim period, when SSP applied for up to eight weeks, as from April 1986, this scheme now covers up to twenty-eight weeks in any tax year.

There are various ways SSP may be paid in relation to any other sick pay arrangements the employer has provided. It may be paid in addition to or instead of normal pay, or an equivalent sum is deducted from that pay.

DSS officers may inspect sickness records at any time, and these must be kept available for at least three years after the end of the tax year in which any particular absence occurs.

A staff sickness record card may be used by an administration manager to record the details of any incapacity of one day or more. The period recorded may include every day of the week, including non-working days such as Sunday, but an employer can opt for other arrangements. The employee signs each entry, confirming the details, or, alternatively, the employee submits form SC1 for an absence of between four and seven days. Any incapacity lasting four or more consecutive days forms a period of incapacity for work (PIW). If in a period of not more than fifty-six days after the finish of one PIW there is another PIW, the two PIWs are treated as 'linked' because of the short gap (and not because of any connection between the reasons for incapacity). SSP is only paid for PIWs, not for the first three 'waiting days', and the maximum entitlement to SSP in any one PIW or series of linked PIWs is 196 days, that is, twenty-eight weeks. After the first seven days of sickness absence, during which self-certification suffices and the employee may not have consulted a doctor, if the sickness absence continues, the employer may require (and a doctor will issue without charge), a Medical Statement Form Med 3. Such a supporting statement is usually supplied by the employee's GP, but may come from a hospital.

A most helpful, clear and readable source of information on the subject is provided in NI227 'Employer's Guide to Statutory Sick Pay'. This makes the point that the employer decides if SSP is payable, decides rules about notification of sickness and decides what counts as evidence of incapacity, making the point that certificates could come from osteopaths, chiropractors, herbalists, and others, as well as from doctors. Examples are given of types of sickness absence with and without linking.

Form Med 3 states that the patient has been examined by a doctor on that day or the day before and has been advised whether or not to refrain from work. Form Med 3 is provided by the doctor who at the time has clinical responsibility for the patient. It cannot be issued on the basis of a telephone call or request from a relative, the patient must personally see the doctor. This reduces the risk of fraud. A doctor who issues Med 3 without seeing the patient is liable to disciplinary measures by the General Medical Council. If another statement is asked for to replace the original which has been lost, it must be clearly marked 'Duplicate'.

In certain circumstances, a Form Med 5 special statement may be issued by a doctor without examination, advising the patient to refrain from work for up to one month on the basis of a written report from another doctor at a hospital, place of employment or other institution, but not from a partner in the same general practice and written not more than one month earlier. Form Med 5 is also used when a patient requires a statement for a past period during which she/he saw the doctor, but no statement was issued.

In 1976, DHSS NI certificates became statements with more flexible rules. In consequence, the numbers issued per year fell from 40 million to 25 million. (On the old certificate, the doctor certified incapacity for work rather than advising the patient to refrain from work). In law, doctors are neither required to issue a statement for a period of incapacity for work which lasts seven days or less, nor for the first seven days of a longer period.

A statement can be 'closed' or 'open'. A 'closed' statement covers the total period of absence. The maximum period covered by a closed statement, from the date of the doctor's examination to the predicted date of recovery, increased from seven to fourteen days (50 per cent of the eleven million claims for benefit each year are for two weeks or less, and 25 per cent were for less than one week). The individual may return to work before the date given on a closed statement, but must notify the local Social Security office. Another statement will be required if the expected improvement does not occur.

The maximum period covered by an 'open' statement also increased, for a first statement to six months, and for subsequent statements for any forward period. If the patient is unlikely to work again, the statement may be 'until further notice'. Once an open statement has been issued, a closed statement must subsequently be issued before return to work. Where the patient has recovered from an illness before seeing the GP, the doctor may write 'unspecified' in the space for diagnosis.

Employers' fears that the changes in statement procedure, that is, increasing the period for which a statement is not required to seven days, would be a 'charter for malingerers' have not been fulfilled. Accurate diagnosis should be given but in the past, there was provision for the doctor not wanting to alarm a patient to give a less precise 'vague' diagnosis on the Form Med. 3 statement he or she gives the patient, and to notify the Divisional/Regional Medical Officer of the DSS separately (on Form Med 6) of the true diagnosis, together with an indication of the prognosis. The DSS would then know not to call the patient early for special examination as might otherwise happen when the diagnosis is not clear. The DSS special examination and monitoring system is involved in cases of unusually prolonged absence, vague or changing diagnosis, or if the patient appears to be consulting different doctors. Initially, enquiries are addressed to the issuing GP and the patient may be called for examination. Benefit can be withheld until the patient has been visited by a local officer of the DSS, or referred to the Regional Medical Service for a second opinion. The claim is then submitted to the Insurance Officer who decides entitlement.

GPs are sometimes criticised by employers for giving out medical statements too readily, but the employers have not appreciated the difficulties – that it is often largely what the patient says to the doctor about symptoms like headaches, which cannot be proved or disapproved, plus the patient's attitude to work, which decides whether or not a Form Med 3 is issued. Thus, to a certain extent, it is the patients who decide their own fitness to work. For a doctor to imply the patient who feels unable to work because of, say, a headache, is lying threatens the doctor/patient relationship. As a result not just that patient, but the whole family could leave the practice, thus reducing the doctor's income, which is based on the total number of patients looked after. The doctor would not give in to such pressure but in addition the family doctor cannot be expected to know in detail the work each patient does, and is reliant on what the patient says about the job.

To summarise, SSP is not paid for the first absence of three working days – the 'qualifying' or 'waiting' days. If sickness absence is for seven days or less there is no need for a medical statement, as self-certification covers an absence of up to seven days, an extension from the time when many employers did not require a certificate for an absence of one, two or three days, relying on employees to inform their manager or supervisor verbally. If such absences are frequently taken, employers may suspect the validity of the reasons given. They used to refer to such days as 'the easy three' and now would refer to the 'easy seven'. An employer could insist upon an employee with such a record obtaining a private certificate from his/her GP for any future absence, however short: their reasoning is that a person is less likely to lie to or try to deceive their GP than their manager or supervisor. However, doctors outside as well as inside occupational medicine do not want to have a policing role.

As shown in the late Dr Peter Taylor's chapter in Professor Schilling's textbook on occupational medicine for doctors, there are many factors apart from the illness or injury itself which may contribute to why a person stays away from work (see Table 6).

Table 6   Factors that influence sickness absence

National/regional	Organisation/ Department	Personal factors
Geography	Type and size	Sex, age and marital status
Race	Management attitude	Occupation and length of service
Season	Supervisory quality	Working hours and wage rates
Health Service	Personnel policy	Job satisfaction
Insurance benefits	Sick pay scheme	Length and complexity of journey to and from work
Pension age	Working conditions	Medical conditions
Epidemics	Medical Service	Family responsibilities
State of economy	Labour turnover	Personality

Some GPs would prefer not to have to provide social security and SSP statements, both because of the time involved, and because they consider it is not their role to adjudicate on fitness to work. They were also opposed to the previous practice whereby, although it was headed 'for social security purposes only', the patient sent the statement to the employer, who took a photocopy and then forwarded the form to the employee's local social security office. This practice was contrary to the policy of the conference of Local Medical Committees and of the BMA, but the DSS Guidance Notes to doctors on the issue of statements pointed out that patients may, if they chose, show it to their employers before forwarding it to the Social Security office.

Some doctors would prefer that a separate, private certificate should be provided for the employer on the following lines:

This is to certify that in my opinion ....
of ....
is *able*   to attend *work*
   unable

                    Signed ....
                    Date ....

They consider the employer should not know the diagnosis, and should only have corroboration that the employee has been advised to refrain from work. The doctor may charge for a private certificate.

In some companies, any medical statement is sent direct to the medical department so that only the company doctor or nurse know the diagnosis (sometimes, indeed, it may be only they who understand the medical terms used). They then tell management who is away and can give a view for how long an absence is likely to last – something management need to know – the all-important prognosis.

Monitoring sickness absence is never easy; especially difficult are short-term sickness absences of up to one week which will be self-certificated and where employees will not necessarily have consulted a doctor. Clearly if such absences occur frequently, it is in the employee's best interests to get a medical opinion, and this is the line a caring employer will take, telling the member of staff to do so or arranging for him or her to see the company nurse or doctor. The DSS describes a scheme by which, if an employer has doubts about the absences and there have been four of between four and seven days in twelve months, the DSS will write, if correctly advised, to the employee, asking that person to visit his or her GP the next time he or she is unwell. The GP will be asked to provide a confidential report to the Regional Medical Services with the consent of the employee. If, on the fifth occasion of sickness absence, the employer doubts that the incapacity is genuine, the SSP payment should be suspended and the employee asked to take the matter up with the DSS. The Regional Medical Service will be asked to give an independent opinion which will be passed to the adjudicating officer, who will decide whether the employee is incapable of work.

Dismissal or termination of contract, though usually for an offence or incompetence, can be justified on the grounds of ill health, either for one prolonged illness or frequent short absences. In prolonged illness, if medical opinion is that the employee will not sufficiently recover in the foreseeable future to return to his or her previous work, it can justifiably be said that the contract of employment has been frustrated and is at an end. Before taking this step, an employer must consult fully with both the employee and his or her doctors. There may be reluctance on the part of the employer to raise this subject with an employee who is ill, the fear being that the resultant distress could make the condition worse.

In a large organisation, suitable alternative work may be possible. Of crucial importance is how long the employer can reasonably be expected to do without the employee. This partly depends on the size of the company and on the employee's role. The absence of one person, whatever his or her job, is more disruptive in a small organisation than in a large company. If the job is a key one, a replacement may need to be brought in early.

When a claim of unfair dismissal on grounds of ill health is taken to an Industrial Tribunal, the tribunal will determine whether the health conditions of employment which are imposed and enforced by the employer are fair and reasonable conditions for the particular job.

Though the outcome of resignation and retirement are the same, these terms have very different meanings as regards finance and future employment prospects, as explained in Chapter 9.

With retirement, a pension starts immediately; with resignation the pension is deferred to normal retirement age. Early retirement on health grounds provides an immediate pension either based on years of service or made up to normal retirement age, but future employment, even on a part-time basis, is then more difficult to obtain. The whole point of such early retirement is, in any case, that the person is likely to remain unfit for the sort of work he or she was doing. Indeed, terms for early retirement may include a clause that, should recovery occur, the former employer will have the option of having the person back to work again (though such an option is rarely taken up in practice).

A few employees take advantage of paid sickness absence by 'going sick' for each and every minor or trivial ailment. Repeated, unpredictable absences, which are disruptive and prevent efficient work and forward planning, have an adverse 'knock-on' effect on other staff, and interfere with the smooth running of a company, are grounds for termination of contract because of ill health, particularly in a small company or where the individual holds a key job.

Termination of contract for repeated STSA must follow a standard procedure with an initial, witnessed, formal verbal warning, or a written warning that includes the likely consequence of continuing absences. If absences continue, a final written warning is issued, stating that if this level of STSA continues, suspension, dismissal or some other action is likely. With continued absences, provided the contract of employment makes provision for this, the employee may be suspended, transferred or dismissed. Detailed records of each step must be made. Obviously, at an early stage an attempt should be made to find out why so much STSA is needed. The opinion of the company doctor is helpful. He or she may examine the employee and/or write to his or her GP, under the terms of the Access to Medical Reports Act procedures.

The company doctor's role is advisory and his or her opinion will be confined to the medical fitness or unfitness of the individual to carry out the duties for which he or she is employed. Decisions about the employee's future employment are made by management guided by this advice. When there is repeated excessive STSA, disciplinary measures – as for any absence from work – rest solely with management.

LENGTH OF ABSENCE

Management would be helped to manage if it could be provided with lists of likely time away from work for a wide range of operations and illnesses. Some conditions, like colds, flu, tonsillitis, and gastroenteritis and other acute infections, are fairly predictable, but the time taken to recover from many illnesses varies widely. It is much more difficult to gauge when a remission will occur in a chronic disorder or how long it will last, or to predict whether treatment will halt possibly progressive conditions such as cancer.

Subjective recovery from any operation, that is, when the patient feels completely back to normal, is also very variable. Much depends on the patient's own feelings, although to a certain extent he or she goes by what their doctor says.

So there are no hard and fast rules about expected length of absence with a particular condition, but certain guidelines are possible about when further enquiry should reasonably be made, when a condition causing sickness absence appears to be dragging on (see Table 7). The Statutory Sick Pay DSS book gives guidance on the more common, vague and less serious conditions, making the point that in DSS experience, the sickness absence situation is usually clearer when the diagnosis is of a grave disorder like cancer (though

Table 7 Duration of sickness absence

Description of illness or diagnosis	Control appropriate by (months)
Addiction (drugs or alcohol)	3
Anaemia (other than in pregnancy)	2
Arthritis (unspecified), rheumatism	2
Back and spinal disorders (PID, sciatica, spondylitis)	3
Concussion	2
Fractures (Colles, clavicle, ribs, fingers, toes)	3
Haemorrhage (stomach or bowel) haematemesis, melaena	3
Headache, migraine	2
Hernia	2
Inflammation and swelling	2
Joint disorders other than arthritis and rheumatism	3
Kidney and bladder disorders, cystitis, UTI (urinary tract infection)	3
Menstrual disorders, menorrhagia, D&C	2
Miscellaneous symptoms and diagnoses – anorexia, debility, fainting, giddiness, insomnia, investigation, NYD (not yet diagnosed), obesity, observation, tachycardia	2
Mouth and throat disorders	2
NAD (no abnormality discovered)	Immediate
Nervous illness	3
Post-natal conditions	3
Respiratory illness – cold, coryza, URTI, influenza	2
– bronchitis, asthma	3
Skin conditions, dermatitis, eczema	2
Sprains, strains, bruises	2
Ulcers – peptic, gastric, duodenal	3
– varicose	3
– corneal	3
Wounds, cuts, lacerations, abrasions, burns, blisters, splinters, foreign bodies (FBs) }	2

my experience is that prognosis is seldom easy in any significant condition).

Dr David Owen, when Minister of Health, quoted a study of six hospitals which found the length of stay in hospital varied with the specialist concerned. After a hernia repair, the range was two to twelve days; after appendicectomy, three to ten days; tonsillectomy, one to five days; hysterectomy, three to eighteen days, and after a heart attack, ten to thirty-six days. The range is now not so great but there still are significant differences.

In the past, extended bed rest after an operation was usual. Nowadays, after, say, straightforward removal of the appendix, the patient starts to get up next morning and leaves hospital a few days later; the stitches are taken out at home by the GP or community nurse. After an uncomplicated heart attack, instead of prolonged bed rest the patient is up in a chair after a few days and home within ten days. Even after a hip replacement with an artificial joint, the patient is walking again in the ward within forty-eight hours, blissfully free of pain. For any particular illness there is considerable variation in the length of sickness absence. No two cases even of the same condition are identical. Reasons why the length of sickness absence for the same illness or operation varies from person to person, include motivation and the type of work performed. For example, a manual worker may need sick leave for a broken toe, whereas a clerk working near home may not.

A male machine operator developed a groin hernia and was referred to a surgeon. He claimed industrial injuries benefit and was put on the waiting list for an operation. He was given sedentary work until admitted to hospital for the operation; he went home a week after the operation, was told to avoid heavy lifting for six months and returned to work after eight weeks, to a sedentary job. A self-employed grocer of similar age developed the same type of hernia after unloading his van. He elected to have the hernia repaired privately and was working normally three weeks after the operation. In neither case did the hernia recur.

When employees used to having access to a nurse at work are relocated to a building without a nurse, sickness absence may increase. This is not just STSA, though an 'off-colour' employee who knows there will be an expert on hand to visit for an opinion and treatment, is more likely, at least in theory, to come to work. Many serious illnesses come on gradually, and employees who would have visited the nurse, may in his or her absence at first carry on. By the time they see their GPs, the condition will have reached a more advanced stage, may need more radical treatment, and may involve longer sickness absence. An example is appendicitis where, if early symptoms are ignored, the appendix may burst and the condition of the patient deteriorate.

Apart from early diagnosis, other factors which influence the rate of recovery from an operation or illness include:

Age – a younger patient usually recovers more quickly and heals faster than an older one.

State of general health – the fit and lean recover more quickly than the unfit or obese.

Any complicating factor – a bad wound infection; an underlying general disorder such as diabetes; smoking cigarettes, a drink problem. Easy to overlook are the after-effects of the anaesthetic.

The 75 000 hernia operations performed each year in the United Kingdom account for about 3500 000 lost working days – four to seven days in hospital and two to three months off work. In Canada, those who have this operation are expected to reach their previous level of activity in four weeks. Many Canadians, unlike British workers, receive only state sickness benefit and do not have their pay made up by the employer.

A survey in the United Kingdom showed hernia recurrence unrelated to early return to work and mostly occurring in sedentary workers, presumably because manual workers, because of their work, are basically fitter and have better developed muscles.

Return to work after an operation may follow a standard pattern. On discharge from hospital the care of the patient passes to the GP but usually, at the time of discharge, a 'follow-up' appointment is made to see the surgeon or a member of the surgical team at an out-patient clinic to check that all is well, the scar fully healed, and so on. This appointment is commonly four to six weeks later and the patient may think she or he must not return to work until having been back to the hospital: in the meantime, the GP may not pass the patient as fit for work. In any case, the person concerned may not want to hurry back to work, particularly if the job involves physical work. Return to work thus awaits the hospital appointment.

One survey showed that the recurrence of hernia in miners on the coalface who went back to normal work one month after the repair, was no more likely than when three months' sick leave was followed by a period of light work. Complications can of course arise during or after any operation and delay recovery. On the other hand, one person who had an operation after many years' good health and an impeccable health record, felt it justifiable to stretch out his recovery just to make up for all the years without sick leave!

NOTIFICATION OF ILLNESS

When a member of staff is ill and needs sick leave, it is understandable that the employer wants to know as soon as possible that this person is not coming to work that day.

Equally important to management is the prognosis – how long will the employee be away from work, and what are the chances of further episodes of the same illness? The diagnosis is a guide to the likely length of sickness absence, but it is debatable how much the employer should know about the illness. In self-certificated sick leave, the reason is given by the employee, and this covers up to seven days away.

The sick employee who stays at home is required to ring in and say so on the first morning away, or if in hospital or too ill to phone, a member of the hospital staff, relation or friend should do so. It is usual practice for the employee to send the Department of Social Security (DSS) Medical Statement, usually form Med 3, issued by the GP or hospital doctor, to the company on day eight of sickness absence. The form used to be marked 'For National Insurance Purposes Only'; now SSP is included in this phrase. When the employer requires a medical certificate to support absence, the employee may choose to ask for a 'private' certificate, where the diagnosis is not disclosed.

Sometimes the outlook is clearcut. For example, if the illness is said to be influenza, the otherwise healthy person should be back at work up to a fortnight later. What is apparent is a misuse or a misunderstanding of what flu is. Many a sickness absence record card is peppered with absences for flu, often several in the same year, each for anything from one to seven days – clearly a nonsense! With a more serious illness, or when every sickness absence is prolonged, being measured in months rather than weeks, or when the diagnosis is not straightforward or cannot be made at all, or if an open-ended certificate has been issued, management needs to know the position, and in particular, when they can expect that person to be back working efficiently. In their efforts to find out, they could cross ethical boundaries of individual privacy and medical confidentiality.

Before any enquiry is addressed to a doctor about a sick employee, that employee's informed consent in writing must be sought in line with the Access to Medical Reports Act. There can be reluctance to seek this consent because it is felt to be the wrong time; because consent is voluntary and the employee could refuse; out of concern at being thought insensitive, or because the mere act of asking could make the condition worse, especially if it is a psychiatric disorder.

For instance, if a member of the management team visits an employee in hospital, it would be wrong to ask probing questions of the hospital staff at that time, relating to cause, progress and prognosis.

There is increasing sensitivity about disclosure of personal information to anyone not strictly entitled to it. In the interests of all, careful protection of confidentiality is paramount. The BMA's handbook on medical ethics states: 'It is a doctor's duty strictly to observe the rule of professional secrecy by refraining from disclosing voluntarily to any third party, information which he has learned directly or indirectly in his professional relationship with the patient.' Exceptions include when the patient gives consent; when the information is required by law; when it is given in confidence to a relative, and when the community interest overrides the doctor's duty to his or her patient.

Confidential medical reports should not pass through many hands en route to a personnel file. A doctor who telephoned one personnel department about a report he had sent was shaken to hear a clerk call across the office 'Anyone seen the file of Mr X – that man with depression?' Normally a GP or hospital doctor will communicate medical information only to another doctor, that is, the company doctor.

Preservation of confidentiality is addressed under data protection legislation; after much debate about the putting of personal medical information onto a computer, the guarding against unauthorised access, and whether, for example, the person concerned should be entitled to a copy of all information so recorded, a patient is now entitled to see any personal notes or other records kept on computer, but not on hand-written or typed files. The term 'computer' includes a word processor.

Occasionally the company doctor is requested to ring a GP or hospital doctor to save time, but even if the doctor at the other end of the line has the signed consent form, how is he or she to know for sure it is the company doctor speaking? There are times when a letter followed by a phone call is needed, but it is usually best to await a written reply and avoid any misunderstanding. The Access to Medical Reports Act applies to the spoken as well as to the written word.

The employee's consent is given for a medical report to be issued to the company doctor 'so that he/she may advise the company'.

Inclusion of this phrase makes the purpose of the report clear to both the employee and the doctor issuing the report. The form is dated and should be used at once. The original should be sent with the letter of enquiry, not a photocopy; but a photocopy should be held on the medical file. If later on a follow-up progress report about the same condition is needed, a fresh consent form must be used. The reply is 'doctor to doctor', and the company doctor then comments to management on medical information which is pertinent to the work situation, avoiding clinical detail and not straying into domestic issues disclosed by the doctor which are no business of the company.

For a logical letter of enquiry and a pertinent reply, the company doctor needs detailed background information, so the letter from the personnel manager to him or her must include full details of the employee and why management are concerned. The details required include date of birth, date of joining the company, brief job description, standard of performance, progress in the firm, sickness absence record and so on, and can be issued as a first contact form as shown earlier in this book.

Then, the company doctor can write to the GP or specialist a standard letter of enquiry which also contains particular facts about the patient. The letter might read like this:

I am writing to you in my capacity as the company doctor and should be grateful for your help with the above-named patient of yours who is employed by this company.

The Personnel Manager is concerned about his/her health and is seeking my opinion and guidance.

To assist me in advising the Personnel Manager, I shall be grateful if you will send me a report from your records on his/her health. It will also be helpful to have your view about the further period of sick leave he/she is likely to need.

I enclose a signed Form of Consent, authorising you to give this information. In compliance with the Access to Medical Reports Act 1988, your patient has today been informed of our action.

The company will be pleased to pay a fee for this report. Please send a separate account with your report for me to forward to the appropriate office.

With many thanks for your help.
Yours sincerely,

One copy of the enquiry letter goes on file and a further copy goes into the diary for four weeks ahead (one week beyond the twenty-one days allowed in the AMR Act), when if there has been no reply, a chasing enquiry is made by telephone. From the time of the original management enquiry, a report from the company doctor may not be received for four to six weeks.

Neither GPs nor specialists are obliged, under the terms of service in the National Health Service, to provide medical reports to companies or their medical advisers, or to do so free of charge, and although sometimes, with consent, such information is provided without any question of payment, they can claim a fee. The amount of the fee should vary with the complexity of the report; the time it takes to prepare; whether a special medical examination of the employee was necessary, or whether the report is made from existing records; whether a statement of medical fact is required, or fact plus opinion – for example, the prognosis. Given the facts the company doctor will be able to provide this.

The British Medical Association (BMA) issues guidelines, reviewed each spring, on a range of fees for various services which doctors, particularly GPs, may provide. These include a range for medical reports to third parties.

The doctor issuing the report to an employer's company doctor is not bound by these recommendations, but it is reasonable when seeking a report to offer the doctor a fee in that range trusting this will be acceptable. The doctor then has the option of negotiating if it is not. The quantity and quality of a report varies a lot. In general, a consultant's report is more complicated than a report from a GP. This is far from invariable but may be why, in practice, a specialist is offered more than the GP; although if the member of staff has been a private patient I think there is no need to offer a fee at all.

Where a medical examination is asked for, the fee is considerably higher; again there are guidelines. The guideline fee for a pre-employment medical examination is slightly lower than for an insurance medical examination. An insurance medical is usually more detailed and also asks for an opinion on longevity – whether the person to be insured is a 'first-class' life risk, and may include an HIV antibody blood test for the presence of the AIDS virus.

Failure of colleagues and management to communicate with, or to visit, an employee on long-term sick leave can be from embarrassment, from not wanting to be thought to pry, or for fear of catching the illness. Properly briefed, that is, forewarned by letter or a phone call, an employee on sick leave, who may otherwise feel ignored after the initial 'Get Well' card, may welcome a home or hospital visit by a manager or other colleague, and would not think of it as 'snooping'. She or he may get a morale boost from the concern and interest such a visit demonstrates. Such tactful, sympathetic support and keeping in touch, may even shorten sickness absence and will certainly smoothe return to work after a long absence.

REHABILITATION AND RETURN TO WORK

A person who has a serious and progressive illness may say they want to get back to work. The explanation may be the need for a target as a yardstick of recovery, wanting to show that they are doing their best to get better, or they may not appreciate or even be aware of the seriousness of their condition, or may be concerned that their job is in jeopardy, or just be bored. Whatever the reason, if it is practicable, let that person (in liaison with the doctors), hold on to that ambition (within reasonable time constraints), and even try out the job again so that the ambition is achieved; it will not only be psychologically beneficial but also such an attempt then makes early retirement more readily accepted when the subject is broached.

Most people returning to work after sick leave start work on Monday. The first week can be especially stressful and tiring, so it is better, where possible and practicable, after sick leave of four weeks or more, to come back mid-week, so the first working week is shorter. Statistics suggest that Mondays are particularly stressful anyway, and questioning the average person about life events shows that getting up and going to work on a Monday morning is what most people regard as the most stressful event of the week! Death and the start of an illness occur most commonly on Mondays. After a long absence, if there is a commuter journey on crowded rush-hour public transport, the first few days back should, where possible and practicable, be on shortened hours, 10 a.m. to 4 p.m., rather than 9 a.m. to 5 p.m., but still include normal breaks.

After mental illness, where there will have been no visible sign, nothing to show anything was wrong – no rash, plaster cast or scar, there may be smouldering resentment against the employee by colleagues who had that person's work added to their own, or from the manager who has paid 'the temp' out of profits. To a sensitive person, any hint of this prolongs the transition back to normal work. This is more likely to occur if the conscientious person comes back to work too early before full recovery, or works shortened hours for long. Management must anticipate and nip such reactions in the bud.

Occasionally, rehabilitation from mental illness includes, with the employer's agreement, return to work part-time for mornings only, the rest of the day and weekday nights being spent at a hospital. This is part of the treatment and is covered by a medical statement and it is important to be quite clear on this point, as normally in no way should an employee be allowed to resume work when not in possession of a formal 'signing-off' certificate.

During rehabilitation from serious illness, or after surgery, liaison, without any attempt to interfere or intrude, between the employer, via the company doctor, and the employee's doctors can be valuable. It may be months before the individual feels one hundred per cent again, but sick leave may not be needed for all that time. Opportune return to work, modified if necessary, can be arranged by liaison between the specialist and/or GP, who are given details of the person's job, and the company doctor, nurse or manager. This avoids use of the vague phrase 'fit for light duties only'. For some jobs there can be no qualification and a person is either completely fit or not for that work.

The employer gives details of the work and whether the work-station, equipment used, environment, type of chair, and so on can be adapted to allow earlier return to work of some kind, and also, where the job allows, whether phased return to work with initial shortened hours, is possible. Temporary alteration of the job description may make all the difference, not having to lift or bend, for example. Nevertheless, there will be those who 'spin out' sick leave.

Apart from the job, commuting must be considered. In London and other major cities, at peak hours, this can be an ordeal which calls for mental and physical fitness. For example, someone subject to prolapsing piles and needing non-urgent surgery, whilst waiting to go into hospital for the operation, should not (in the meantime) commute standing in the rush hour to a job where they are standing most of the day. However, they could likely manage a sedentary job and commuting outside the rush hours. Unless some liaison occurs, such a person is likely to continue on sick leave whilst on the waiting list and until fully recovered from the operation. So both the journey and the work itself are subjects for the company doctor or nurse or, if neither is employed, for the personnel manager, to discuss with the GP and the employee. The GP may contact the company or the

company the GP. In either case, the informed writ'
employee is required, as per the Access to Medic
the employee must understand why a report is ne
read the report before it is sent.

Alternative employment

In those cases where some aspect of the work itself, or the working
environment, appears injurious to the employee's health, because of
susceptibility to a particular illness and not due to a hazard of the
work or environment, the alternatives are *job transfer*, *resignation*,
early retirement, or *termination of contract*. The same four alternatives apply when commuting causes or aggravates an existing condition, or when there has been protracted sick leave. Alternative
work should always be considered unless the person has a terminal
illness. A suitable, alternative job does not have to be newly-created,
but there may be the opportunity for a sideways move. Ideally, the
job should be of the same grade and pay scale, but where it is felt the
pressure of work contributed to the illness, a lesser job with less
responsibility, and inevitably less pay, may be necessary, or stopping
work altogether (for example, the person with coronary heart disease
who suffers angina when under stress at work, or the manual worker
with recurrent backache).

Even in a large company, it is not easy to arrange a transfer on
health grounds, nor when the request is for a move to a local office,
because of difficulty commuting to a big city. However, an attachment to an office nearer the patient's home can sometimes be set up
temporarily during convalescence. For a permanent transfer on
health grounds, the ailing employee has to be 'sold' to the local
manager, who may be unhappy at taking on what he or she sees as a
liability, who may often be away sick.

A person with angina or an arthritic hip may have increasing
difficulty with a journey which involves much walking or many stairs.
Once at work, though, there may be little or no problem if the work
is sedentary and routine. More controversial are the occasions when
the complaint is that the journey to and from work is too tiring or
aggravates, say, a chronic chest or back condition or a psychiatric
illness. Few people enjoy commuting, and companies which have
difficulty recruiting staff in the centre of a large city, especially
London, may be reluctant to agree to a local move in these circumstances, even if local work is available, especially where an em-

ployee, who initially lived near his or her work, then chose to move house further away, preparing perhaps for retirement, still several years off. The longer journey then affects health and reasonably the GP may suggest early retirement on the grounds of ill health if a local transfer is not possible – but for the employee there may be the option of moving back to be nearer the work place once more.

FREQUENT ABSENTEES

With minor illness or injury (over half the sickness benefits paid each year are for two weeks or less), in many instances it is the individual as much as the doctor who decides fitness for work. As Table 6 showed, many factors influence this decision, apart from the illness itself, and the GP's opinion. These include the type of job, colleagues, and commuting journey, as well as domestic, national and environmental factors. Is the work interesting and responsible, or dull repetitive and boring? Is the working environment pleasant or merely tolerable? Is there a team spirit?

The number of changes on the commuting journey, and the availability of a seat, influence the decision to stay away more than total travelling time or distance. The more changes with chances of missed connections, and so on, the higher the sickness absence. Pocock & Taylor analysed the journeys and absences of 2000 office workers in central London, and found that those whose journey had four or more stages had 20 per cent more time off work than those with three or fewer stages.

When suffering from a minor ailment, a seat on a train with a short walk at either end is obviously more tolerable than having to change from bus to train and train to tube, waiting in a queue in the rain, with the prospect of the same journey all over again in the evening. Also relevant is the time of year: any journey is easier on a dry, spring morning than on a wet winter evening. On the other hand a smooth commuting journey provides space, time to switch roles from home to work in the morning and back again in the evening.

SHORT-TERM SICKNESS ABSENCE (STSA)

In a basically healthy person, once a high STSA pattern for minor illness has been established, it tends to persist. This pattern may be established in childhood, if school is disliked or parents overprotective. Health and attendance records should always be asked about in reference enquiries.

When unemployment is high, STSA falls. Many other factors, such as stress of work, influence STSA; stress can be a spur to some but others fail to cope, become inefficient and ineffective and take days off when it gets too much. Repeated STSA gives the employer cause for concern, but the GP if not consulted, may be unaware of a health problem. Deterioration in work that appears linked to health is another reason for seeking a medical report before considering disciplinary action, terminating service or demoting to a lesser grade. Disturbed nights with a new baby cause daytime tiredness, and may, for a time, affect punctuality and efficiency. Illness of the spouse, especially mental illness, where there are children to be looked after, or illness of the children, will distract from work.

A possibly suspect pattern of STSA is Friday and/or Monday absences, lengthening a weekend off when most staff not involved in weekend shift work are at home.

Whilst all employees are expected to give the reason for STSA, this is often verbal. A deterrent to malingering is the requirement, on a sickness record card, for a statement of the cause of the absence to be signed by the employee on return to work. This is the practice for self-certificated sick leave.

STSA rates are likely to be lower in a small compared with a large department, where others provide cover for essentials. 'One less will make little difference, I won't be missed.' In a department where many staff do similar work, a way to avoid this attitude is to break down large numbers into small groups and foster team spirit, so that no member of that team wants to let the others down by being away, unless it is quite unavoidable.

An effective on-site first aid and nursing service, in which staff have confidence may reduce STSA. With a company nurse available employees know they can get professional advice and that for any serious or recurrent condition they will be referred to their GP.

Early and efficient treatment of minor injury by the first-aider ensures prompt healing.

STSA is higher amongst staff about to leave, 'demob-happy', other than those about to retire, and lower amongst new employees, particularly in a probationary period during which performance, timekeeping and so on will be appraised before an appointment is confirmed, or when initial employment does not qualify for paid sick leave.

High STSA may reflect dissatisfaction and frustration at work rather than ill health; if so, control and remedial action lies with management; overpromotion, underpromotion, poor communications within a department, and boring dead-end jobs are examples of likely issues. If this signal is ignored at the early stage, when morale is evidently drooping, frank physical, mental or behavioural disorders may follow. One way to provide job satisfaction in the form of job enrichment is described by Bill Walsh in *Education and Training*. He recommends 'client identity', a relationship with 'some identified and meaningful entity', such as customers or another department, 'which invests what the clerks do with a responsibility to that entity'. This provides the opportunity for readily recognisable good service, very different from the service given to the 'amorphous' organisation. Creation of more satisfying office jobs is an objective of the Work Research Unit of the Department of Employment, which provides a free advisory service (Steel House, 11 Tothill Street, London SW1H 9LN).

A few companies offer additional leave to those who take no sick leave, but such a scheme may be counterproductive. It is obviously better for someone with an infectious illness, such as tonsillitis, to stay at home, rather than come to work and pass on the illness to other staff who might otherwise have remained well. Also, someone who feels ill is unlikely to work as efficiently as when well. It has been shown that middle-aged men with many years 'nil' sickness absence are more liable to a heart attack than those with average sickness absence. Some STSA may thus protect against more serious illness. Those with low or nil sickness absence should perhaps occasionally be reminded that STSA for 'minor' illness is sometimes necessary for long-term health and that not everyone, though needed, is absolutely indispensable all the time.

MALINGERING

Malingering was neatly defined in the *Lancet* by Drs Miller and Cartlidge as 'all forms of fraud relating to matters of health'. In an occupational setting, deliberate invention of symptoms of illness in order to avoid work is uncommon, but can occur. (Some people repeatedly seek hospital treatment by manufacturing symptoms, the so-called 'Munchausen Syndrome' after Baron Munchausen, a legendary boaster and liar.) Malingering can be a means of expressing dissatisfaction with the conditions of work, and can be difficult to prove, because, more often than not, it is not a case of pure invention but an exaggeration of an existing symptom, or, particularly where there are permanent X-ray changes, a claim that a past symptom has recurred, or, in a compensation case, the symptoms are prolonged, clearing only after the case is settled. The deception may be conscious, partly conscious or subconscious. According to Dr Naish, writing in the *Lancet*, nearly all forms of deception are now accepted by the medical profession as a form of illness.

In his book, *Talking Sense*, the late Dr Richard Asher defines malingering as the invention, production or encouragement of illness for a deliberate end. He advises caution in making this diagnosis; too many apparent examples turn out to be cases of organic disease. On a lighter note, he tells a story against himself about his mistaken diagnosis of malingering by his small daughter, who, as he dressed her to take her unwillingly for a walk, behaved badly, cried, walked with a curious, scissors, gait and frequently fell to the left. Such behaviour persisted during the walk and all this Dr Asher put down to malingering until at bathtime his wife exclaimed, 'Do you realise you put both her legs through the same hole in her knickers!' Dr Asher classifies malingering by motives of fear or desire – for compensation, early retirement, revenge, and so on, or for escape. At the medical borders of malingering are hysteria and hypochondriasis. Cases of hysteria, more often than not, turn out to have a physical cause. A hypochondriac constantly fears the worst as the explanation for a symptom, and as soon as one symptom is explained and clears up, rather than be reassured, he or she will latch on to something else. The malingerer, on the other hand, tends to persist with one particular symptom, especially one which cannot be disproved.

14 Accidents at Work

Chance, coincidence of time and place, and the environment, are some of the reasons why accidents happen; others are poor design and/or maintenance, mistakes and plain carelessness.

The concept of accident proneness, developed in the USA some decades ago, has fallen into discredit somewhat, but some people do seem to suffer more accidents than others. There may be a background of chronic mental illness. There is a link between accidents and alcohol, one reason many American companies are 'dry' during working hours and insist that alcohol and work are kept separate. In the United Kingdom at least one company breathalysed shift staff on a voluntary basis on their arrival for work, and only those who 'passed' were allowed to start work. Fatal injury from an accident is the fifth commonest cause of death in England and Wales after heart disease, cancer, strokes and lung disorders.

Each year 18 000 people die from accidents; 300 000 more suffer severe injuries and 5 million, minor injury. Many are the result of road traffic and domestic accidents, but accidents at work account for over 600 deaths each year, and 400 000 work accidents are bad enough to cause more than three days' absence from work. Most of the fatal accidents happen on building sites and down mines. By comparison, offices are much less hazardous, but it is wrong to assume that offices are always totally safe. Each year about 25 fatal and 5000 serious injuries from accidents occur in offices.

Serious injuries are those which cause over three days' sickness absence, and so by law have to be reported. In addition, there are very many accidents where less sick leave or none at all is needed, but which nevertheless cause distress and pain. These very often could have been prevented and need not have happened. Office accidents follow a broadly similar pattern to accidents in industry and result from the same basic causes.

To improve on these figures the causes, treatment and prevention of accidents must be constantly reviewed. The risk of accident was one reason for passing the OSRP Act, but where this places responsibility for safety mainly on management, HASWA, as outlined earlier, also highlights the responsibility of employees to work safely and to report potential hazards to management. In essence, HASWA states that employees must take reasonable care for the health and safety of themselves and of other persons who may be affected by their acts or omissions at work. HASWA also says that the employee must not intentionally or recklessly interfere with or misuse anything provided in the interests of health, safety or welfare. This covers, for example, the unauthorised removal of a guard covering a moving part of a machine.

ACCIDENTS IN THE OFFICE

The commonest office accident (around 50 per cent), is falls, both falls on stairs and, almost equal in number, falls on the level in offices and corridors. Sometimes shoes are to blame, a badly-worn heel, slippery sole or poorly-fitting sandal; a person wearing high heels is more likely to have a fall than if flat shoes are worn; any heel can catch in a rubber band, carelessly dropped on the floor. Trailing telephone cables, and, as in the home, loops of flex to a desk lamp or computer – a clear case for wire management – are obvious tripping hazards, as are loose mats that move and threadbare carpets or their curling edges.

No amount of non-slip polish on wood or vinyl floors will stop a slip and fall if spillage from an overfull cup of tea or coffee is not immediately wiped up.

Many of the rules for avoiding a fall are common sense. In essence, 'watch your step', look where you are going, do not carry so much that you cannot see your feet. On stairways, hold the handrail, don't read, and don't carry too much. Never run, except in dire emergency; it is safer for you, and everyone else, to walk. It may be quicker to use a chair or stool to reach a high shelf, but don't – take the trouble to get steps with sound non-slip rungs and feet.

Good husbandry prevents many accidents – fix cabinets to the wall; top-heavy filing cabinets may topple over when the top drawer is opened. Don't pile books and papers on top of filing cabinets; don't balance heavy ledgers on narrow shelves; don't leave open the bottom drawer of a desk; avoid clutter in passages and on stairways.

Cuts and puncture wounds are other common office injuries. Paper cuts are painful but superficial. Replace pins with paperclips or staples (embedding a staple in a finger is another office injury). Knives, scissors and guillotines too can cause injury, and must be handled with care and used and stored correctly. All injuries, however minor, must be recorded and treated properly and promptly with the appropriate first aid.

Balancing a tray of beakers full of hot drinks from a vending machine or canteen back to the office, risks an accident, either to the person carrying the tray, who is scalded by spillage, or if the floor is made wet by spillage, to someone else who then slips and falls. This practice should either be banned, or only capped non-spill beakers provided, or a type of tray used with holes for the beakers to prevent sliding.

One joker punched small perforations in the bottom of all the plastic beakers, so wetting clothes, shoes and the floor. Back in the office, the drinks should not be consumed at the workstation if there is any electrical equipment there.

Broken glass must be disposed of properly, carefully wrapped, not put loose in a waste bin where the office cleaner, emptying the bin later, may suffer a cut hand.

Any electric shock can kill – there is no such thing as a slight shock – but if the shock does not kill, then there is often no injury, so electric shocks are under-reported. When a mains operated machine develops a fault, the rule must be: don't tinker, switch off, report the fault and wait for an electrician to put things right.

Every office must hold and regularly circulate clear instructions to staff about what to do in the event of a fire or bomb alert. Everyone must be familiar with alarm signals (tested weekly), evacuation procedure and assembly points outside the building. A fire drill clearing the building must be practised at least twice a year. It is surprising how many people gather round the main entrance; if a bomb went off inside they would be hit by falling masonry or flying glass, or at the very least be in the way of the rescue and fire-fighting services. Fire-fighting equipment in the office must be checked once a year and certified in working order. Staff must know how to use the equipment. There should be a fire marshal for each office to co-ordinate these activities and ensure all staff and visitors are accounted for. Preventive measures include:

Put all waste paper in waste bin.
Don't use portable electric-bar fires.
Never put ash, matches, cigarette ends in wastepaper bins.
Keep the lids of all containers of flammable liquids tightly closed.
Switch off electrical machines and unplug before leaving at the end
  of the working day.

Practical jokes which involve horse-play, for example, pulling a chair away from behind a person about to sit down, are dangerous and an offence under HASWA.

A cartoon and caption can get over a safety point better than a plain notice, especially if changed frequently. After a while the same notice may be given no more attention than the wallpaper.

ACCIDENTS IN CATERING UNITS

Catering units are higher-risk accident areas than offices, and, a first aid kit should be provided in the catering unit for immediate emergency treatment. Catering staff should know the elements of first aid for the likeliest injuries. Another reason for this provision is that catering units usually open early, an hour or more before the arrival of the rest of the work force, including the regular first aiders, and may stay open later or, for shift workers, throughout the night.

Likely injuries are cuts, burns, scalds, and falls from slipping on a patch of wet floor. There is also the risk of injury from kitchen machinery such as a food mixer, mincer or vegetable slicer. These are covered by the Prescribed Dangerous Machines Order. The order requires that no employee shall work at any machines listed below unless fully instructed in the dangers and precautionary measures, and sufficiently trained or adequately supervised:

Power-driven machines:
  Worm-type mincing
  Rotary knife bowl-type chopping
  Dough brake and mixer
  Food mixer used with attachments for cutting or crumbing
  Pie- and tart-making
  Vegetable-slicing

Machines power-driven or not:
  Circular knife slicing
  Potato-chipping

SAFETY SIGNS

Safety signs throughout the European community are standard (see Table 8). This does not mean that member states are told where to

Table 8   Standard European safety signs

Shape of Sign	Colour of Sign	Background colour	Wording/Symbol colour	Meaning
Circle	Red	White	Black	Mandatory
Circle	Red edge & transverse bar	White	Black	Prohibition
Triangle	Yellow (black-edged)	Black	Black	Caution
Square or rectangle	Blue	White	White	Mandatory information
Square or rectangle	Green	White	White	First aid areas Emergency routes

put safety signs, but when such signs are posted they are identical in each country, in terms of shape and colour of both the sign or theme and the wording or symbol, which is kept as simple as possible.

ACCIDENT REPORTING PROCEDURES

An accident at work, no matter how trivial any resulting injury appears at the time, must be reported in the Accident Book (usually B1.510), available from HM Stationery Office. This book must be provided under the Social Security (Claims/Payment) Regulation 1979, in a work place where ten or more people normally work. B1.510 must be readily available at all reasonable times to an employee who has been injured at work and to anyone acting for that person. When the book is full, it must be kept for at least three years from the date of the last entry. Each entry is official notification to the employer of an accident. The entry can either be made by the injured person or by someone else on his or her behalf. The signed and dated entry must include the full name, home address and occupation of the injured person; the date, time, location and apparent cause of the accident; the work being done at the time; and the nature of the injury; plus the same personal details of whoever else makes the entry for the injured person. The employer may also record accidents on computer, or keep photocopies of the F2508 form used for reporting an accident to the relevant authority.

All accidents should also be reported to the company safety adviser, the more serious immediately by telephone, so that the circumstances can be investigated and accurate statistics kept for analysis.

The requirements for the reporting to the Health and Safety Executive (HSE) of certain, defined injuries from accidents at work, and of certain diseases relating to the type of work and certain events, termed 'dangerous occurrences', at the work place, are contained in the Reporting of Injuries, Diseases and Dangerous Occurrences Regulations 1985 (RIDDOR). HSE publishes a guide (HSR 23), obtainable from HM Stationery Offices, which also lists the HSE Offices where such reports are to be sent. As regards injuries, these regulations apply not only to accidents to employees, but also to members of the public when the accident involved the company's activities and/or occurred on company premises. The employer must give employees clear guidance about reporting accidents in which they are involved to the person responsible for notifying the enforcement authority.

Also, the company's insurers must be told in case there is an employer's liability (EL) claim. The premises manager needs to know if there is any possibility of a fault in the fabric of the building, or in a fixture or fitting; likewise the engineers may need to be notified.

The injured person will be shaken and so may be temporarily confused. The first-aider, company nurse or doctor who provides treatment should write their own clear confidential notes (trust in an occupational health service is linked to staff understanding this). The notes will be a useful *aide-mémoire* if there is, much later, a claim for damages. Any report then is subject to the written consent of the injured person and will be confined to a description of the injury and treatment given. The first-aider, nurse or doctor should neither supervise nor complete the B1.510 Accident Book entry at the time of the accident.

The employer must, within seven days, notify the Environmental Health Officer at the local office of the enforcement authority on Form F2508 of an accident at work resulting in injuries which causes incapacity for work, or the need for lighter duties, for more than three consecutive days, including rest days, the weekend, but not the day of the accident. Death at the time or within a few days of an accident must be reported by telephone at once, as must a major injury; this includes fracture of skull, spine, pelvis, arm, wrist, leg or ankle. The same applies to amputation of a hand or foot or all or part

of a finger, thumb or toe; penetrating injury to an eye, chemical or
hot metal burn of an eye or loss of sight of an eye for any other
reason; also electric shock causing an injury, including a burn, which
requires immediate medical treatment or causes loss of consciousness. This latter category must also be reported if it is due to lack of
oxygen or absorption of any substance by inhalation, ingestion or
through the skin, or if any of these cause acute illness needing
medical treatment. This latter category also applies if it resulted from
exposure to a pathogen or infected material. Also included is anyone
admitted to hospital immediately after an accident and detained for
twenty-four or more hours. Subsequent death within a year of and
resulting from the accident must be reported in writing. Also to be
reported at once is what is termed a 'dangerous occurrence' which
includes an explosion, or fire, or electrical fault causing work to stop
for more than twenty-four hours.

As regards reportable diseases, the employer has to report on
Form 2508A twenty-eight specified conditions, having been notified
of them by doctors on F Med 3 Statements. As opposed to prescribed
industrial diseases, most of the conditions covered by RIDDOR only
apply to a small number of employees in well-defined jobs. Few could
apply to office staff – one exception is hepatitis B for first-aiders,
company doctors, and nurses, who may have to be engaged in 'work
involving exposure to human blood products or body secretions and
excretions'. Each employer should establish which, if any, of the
listed diseases could apply in the work place from the listed work
activities. Doctors have been urged to use clear standard wording to
describe these diseases on statements but it seems this system is not
working well and may therefore be changed.

Form Med 3 replaced the Medical Certificate in 1976. Section B on
the reverse of this form refers to claims for industrial injury or
prescribed industrial disease, asking for details of employment and
the time and place of the accident. The person who is unable to work
because of the accident or disease obtains such a statement from their
GP, completes parts A, B and C to claim benefit, and sends the form
via the employer to the local office of the DHSS.

THE SAFETY REPRESENTATIVE

The Safety Representative is a member of, and is appointed by, a
recognised trade union, and has worked for the employer for at least

two years, or has had that amount of experience in similar employment. He or she is entitled to paid leave to carry out the functions of a safety representative. These are to investigate potential hazards, dangerous occurrences and the causes of accidents at the work place; to investigate employee complaints relating to health and safety or welfare at work; to represent employees in consultation with the employer on safety matters, and to make representations to management in writing following investigation of complaints; to represent employees in consultation with HSE Inspectors and to receive information from them, and to attend Safety Committee Meetings. Numbers appointed depend on the size of the work force, the size and number of the work places, the shift systems and the potential dangers involved.

Safety representatives have no responsibility in criminal or civil law arising out of their duties.

The common aim of avoiding hazards and so minimising accident risk should unite management and safety representatives to work together. Some companies run in-house training courses for safety representatives; so does the TUC which also recommends union-approved training courses. Safety representatives are also entitled to a reasonable period of paid leave for training. Periodic safety tours, sampling and surveys, are seen as methods of formal inspection which the safety representatives and employers carry out together. The safety representative's role is to ensure that the employer provides a safe work place.

All the above is contained in the Safety Representatives and Safety Committee's Regulations, and amplified in the Code of Practice of HASWA. These provide a framework rather than detail for every situation.

The Safety Officer

The employer may appoint a Safety Officer, a legally-responsible specialist whose role may have previously been included with security or regarded as part of the personnel function. Other sources of safety officers were from the police force or the armed services. There is now an increasing demand for experienced and professional, trained safety officers who will advise on technical aspects of the work as well as on general preventive measures.

The University of Aston in Birmingham has a three-year full-time degree course in Safety and Health; there are also several sandwich, full-time and part-time courses preparing for the examination for the associate membership (AMIISO) and membership (MIISO) corporate grades of the Institute of Industrial Safety Officers (222 Uppingham Road, Leicester, LE5 OQG). These are the most widely recognised safety officer's qualifications.

There is no statutory requirement for an employer to appoint a safety officer nor is there any requirement for registration. The closest that the law comes to such requirements are provisions of the Construction (General Provisions) Regulations 1961, and the Shipbuilding and Ship Repairing Regulations 1960, which require the appointment of supervisors to oversee the respective regulations.

The Safety Committee

Before HASWA, many companies had safety committees made up of representatives from management and the work-force unions. Under HASWA, such a committee, moderate in size and equally representative, must be set up by the employer when at least two safety representatives request this in writing.

The company doctor and safety officer should be ex-officio members of the committee; other relevant specialists advise at particular meetings which are held regularly, under the chairmanship of the manager responsible for Health and Safety.

The terms of reference of the committee are to promote 'the health, safety and welfare of persons at work'. The objectives of the committee are succinctly set out in the Health and Safety Commission's notes of guidance.

One obvious activity is to look, in detail, impartially, at accidents, both statistically overall and at specific incidents individually. The purpose is not to lay blame but to prevent a recurrence and so to consider suitable precautions, make appropriate recommendations for action and set a timetable for these to be completed, and to see this happens.

15 Aspects of Physical Health

PRESCRIBED DISEASES RELATED TO OFFICE WORK

The Social Security (Industrial Diseases) Regulations 1985 and their amendments list over fifty conditions, some with subgroups. Anyone definitely suffering from one of these conditions is entitled to state benefit, but in many cases there are strict qualifying conditions of employment before the condition may be prescribed. These include the time exposed to the hazard and/or working in a particular industry. Most only occur in specific industrial settings not covered by this book, but there are some, to be described in detail, which could occur in a nonindustrial work setting. Once qualification is established, the claimant is entitled to payment in compensation. Also covered by the Regulations for industrial disease benefit is personal injury (rather than accidental injury) caused by the nature of the work, even where the qualification of particular employment listed in the Regulations is not fulfilled.

If the disease causes absence from work, industrial disease benefit is paid from the date of development of the disease for a total of twenty-six weeks. After this time, disablement assessment establishes the amount of disablement benefit to be paid. The latter procedure is also followed when the condition does not cause absence from work. If death is due to a prescribed disease, industrial death benefit will be payable.

Application is made to the Insurance Officer of the DSS who, having established that the necessary conditions are fulfilled, arranges an examination by a medical board. Should there be three rejections by this board, the claimant may appeal against the decision about the medical diagnosis to the Medical Appeals Tribunal. Appeals on the question of non-medical qualification are made to a Local Tribunal and from there to the Commissioner. The appeals procedure has been under review by the Industrial Injuries Advisory Council. The Council also assesses whether a particular disease should be prescribed and if so, on what grounds. The Council is made up of representatives from both sides of industry appointed, after

233

consultation with appropriate authorities, by the Secretary of State for Social Services, who directly appoints the Chairman. Members serve five years and the full Council meets about twice a year.

Following representation, the Secretary of State formally requests the Council to investigate justification of a disease for prescription. The Council acts like a jury, calling on expert opinion as well as on the evidence of people suffering from the particular disease being considered.

Apart from, and sometimes confused with, the Prescribed Industrial Diseases are the sixteen Notifiable Diseases under the Factories Act. Not all of these diseases are work-related and therefore not all of them are also prescribed. The principle behind notification is that urgent action may be necessary to protect the remainder of the work force. Investigation is carried out by Inspectors. Both the doctor involved and the employer (on form 41) required to make a report to the local office of the HSE and to the Area Employment Medical Adviser (Form 43 is used for industrial accidents).

Writer's Cramp

'Cramp of the hand or forearm due to repetitive movements' is one such prescribed industrial disease. 'Writer's cramp' is the best known term used: also 'occupational cramp', 'twister's cramp' and 'craft palsy', titles which suggest the classic symptoms and those who get them. The qualifying occupation is one involving prolonged periods of handwriting, typing or other repetitive finely co-ordinated finger, hand, or arm movements.

Contrary to what one might expect, most sufferers are skilled and experienced, not beginners. The condition comes on gradually, and particularly affects those who must work fast and accurately. At first there is difficulty with hand control when the person is tired at the end of the working day. Mistakes are made. These cause anxiety. A vicious circle develops which perpetuates the condition, especially in a nervous obsessional person.

In established writer's cramp, just trying to write brings on painful hand spasms in that hand, spreading to the whole arm. The limb becomes rigid and painful; writing is impossible; or there are jerky, unco-ordinated movements of the fingers and any writing is illegible. The muscles which have been in spasm continue to ache for a time; this aching may also involve neck muscles. Use the other hand and the condition is quite likely to occur there as well.

Tests show no disease process in the muscles or nerves; usually the disability is confined to one particular act. Other activities with the same muscles, like shaving or sewing, cause no pain or muscle spasm.

Professional writing is the most common occupation affected, but there are several other activities where these cramps may be a problem, including professional piano- or violin-playing. Men are affected more than women.

First-line treatment is to stop work, either completely or that part of it which provokes the cramps. Psychological assessment is followed by psychotherapy and learning techniques to induce relaxation, together with specific exercises for the most efficient and economic use of the muscles involved. The outlook is uncertain. In some there is complete recovery, usually after a series of remissions and relapses lasting several years. In others, the condition remains static and the individual is able, with some difficulty, to continue his or her occupation. Often, unfortunately, the condition is progressive and necessitates a complete change of occupation.

Tenosynovitis

Another prescribed industrial disease which can occur in an office setting is 'traumatic inflammation of the tendons of the hand and/or forearms or of the associated tendon sheaths'. This descriptive title is shortened to peritendinitis crepitans or tenosynovitis. The condition is caused by friction between the tendons and their sheaths due to repetitive, often unaccustomed, use of the wrist and hand.

Superficially similar to occupational cramp, in that work has to be stopped because of pain, the main difference is that whereas with occupational cramp there are no confirmatory clinical signs, here there is swelling in the line of the tendon, and on feeling the area and moving the wrist and hand, a distinctive fine creaking known as 'crepitus' can be felt. Treatment is by resting the arm, in a sling or splint, to allow the inflammation to subside over a period of about three weeks. After this, excessive use of the fingers and thumb should be avoided for a further two months. Tenosynovitis is described in more detail in the section on visual display units.

The beat disorders

Other prescribed diseases possibly relevant to the office setting are those in which the hand is affected by subcutaneous cellulitis (inflammation of the tissues beneath the skin) or where this inflammatory condition or bursitis (inflammation of the lining of the sac that lies between a bony prominence and the skin), affects the hand, knee or elbow. The term for each incorporates the adjective 'beat', thus 'beat hand', 'beat knee', 'beat elbow'. The qualifying condition is manual work causing severe or prolonged external friction or pressure on the hand, elbow or knee. An office cleaner or kitchen porter in a catering unit could be affected. 'Beat hand' usually affects the palm and palmar side of the fingers and thumbs. It is mostly seen in people returning to manual work after a break such as a holiday, when the hands have become relatively soft. There is throbbing pain, redness and warmth. 'Beat knee' similarly occurs in those unaccustomed to kneeling. Repeated minor trauma and wet skin are contributory factors. There is pain and swelling of the lower area of the knee cap where the bursa is situated, or inflammation of the tissues under the skin.

'Beat elbow' may follow a single injury or a repeated minor injuries from leaning with the elbow on a hard surface, as in 'student's elbow'. Here again, there is inflammation with swelling and pain. Cure should follow resting of the affected part for a few weeks. In some cases, as the DSS Notes on the Diagnosis of Occupational Diseases explain, skilled and timely surgical intervention can do much to prevent serious sequelae.

Non-infective dermatitis of external origin

Here the qualifying conditions are 'exposure to dust, liquid or vapour or other external agent capable of irritating the skin'.

Excluded is dermatitis due to infection by bacteria, fungi or parasites, except where the dermatitis becomes secondarily infected. Dermatitis and eczema (Greek: 'to boil over or break out') mean the same thing – inflammation of the skin. There is a sequence of redness, swelling, blisters, weeping, crusting, scaling, splitting or fissuring and thickening (from scratching), with increased pigmentation. Not all the stages need occur and all can clear completely. By no means everything is yet known about eczema. There are many different types of eczema: varicose, seborrhoeic, discoid, pompholyx and flexural for example, each of which has different characteristics.

Sometimes no external factor is involved. This is atopic or *constitutional eczema*, also known in babies and toddlers as infantile eczema. It affects skin creases; is common in childhood when at least

1 per cent of the population of this country is affected; is usually mild, and is linked to asthma and hay fever, or a family history of these conditions or of eczema. Usually this eczema clears completely by adolescence, but in a few people it persists into adult life. Cold weather and stress make this sort of eczema worse.

Other eczemas follow contact with a *skin irritant* – either minimal exposure to a strong irritant like a caustic, alkali or acid, or repeated or prolonged exposure to a weak irritant like a detergent. Both commonly affect the hands when protective precautions have been inadequate. Then there is *allergic contact eczema*, where sensitisation of the skin occurs to one or more specific substances called allergens which include many chemicals, rubber, metal (particularly nickel), dyes and leather. I have never seen allergy to banknotes, but occasionally from coins containing nickel. The substance responsible is identified by patch-testing, in which tiny samples of suspected substances, plus an inert control, are put in a line on the forearm skin for a set time and the reactions compared. A wheal and flare reaction occurs if there is an allergy. Patch-testing must be done by an expert as there are risks involved, one being that the test itself may cause sensitisation.

Sensitisation usually follows skin contact from handling the substance, but can also be due to skin contact from airborne particles, or inhaled substances reaching the skin via the bloodstream. This condition is particularly common on the hands of people involved in 'wet' work like washing-up, when it is aggravated by rubber gloves and occurs under bracelets, rings and watch straps.

A person with contact eczema must avoid the allergen(s) to keep well. Someone with atopic eczema must always be on guard, as their skin can react to a range of substances, such as, pollen. Mixed types of eczema may occur. Any eczema can become secondarily infected and would only then be contagious.

Prevention is all important. Those with dry skin will always be susceptible to eczema, so they especially must keep the skin soft, (emulsifying ointment added to the bathwater helps), use soap sparingly and keep away from anything which can irritate their skin.

If eczema does break out, the general practitioner or skin specialist (dermatologist) will try to identify the cause. There is a range of effective treatments. Eczema sufferers should not hesitate to seek medical advice about the appropriate treatment. This is far better than blindly trying a variety of preparations from the chemist in the hope of finding one that works. In some cases a form of local steroid helps, but whilst steroid lotions, creams and ointments (some of which are now available over the counter instead of on prescription only), can have an almost magical effect, they have their drawbacks if used too strong or for too long. For instance, the skin thins. If the cause has not been found before steroids are prescribed, it becomes much more difficult to make the diagnosis later.

Psoriasis is quite different from eczema and will be dealt with later in this chapter.

HEART AND CIRCULATION DISORDERS

Blood pressure (BP)

Blood pressure reflects the force of heart-beat and the resistance in the arterial tree. Of the two readings, the lower (diastolic) was thought the more significant; now the upper (systolic) is regarded as of equal importance.

BP varies during the day, falls with rest, and increases with mental or physical activity.

That BP and weight should increase with age is no longer believed, nor that women withstand raised BP better than men.

There are no characteristic symptoms of early hypertension, the medical term for raised BP, and in 95 per cent of cases, no known underlying cause. Sometimes, after total unawareness for years, there is a disaster like a heart attack or stroke. My impression is that unrelieved mental tension may contribute to persistent raised BP. In this country, untreated, raised BP reduces the life expectancy of a forty-year-old by up to fifteen years; it contributes to one in eight deaths from heart or kidney disease or from stroke, and increases fivefold the risk of getting these diseases.

One life assurance company reported 'moderately' increased, untreated BP as reducing life expectancy by sixteen years.

In the USA, the Department of Health Education and Welfare believes there are eleven million Americans with undiagnosed hypertension; they urge everyone to visit regularly their doctor for a check, and arrange mobile clinics for BP measurement in shopping centres. Their advertisement reads, 'High Blood Pressure is like a time bomb ticking away, ready to kill or cripple at any moment'.

That it is so simple to discover raised BP, that raised BP can readily and effectively be treated, and that to do this saves lives, are very good reasons to recommend five-yearly blood-pressure screening for everyone.

What is still not clear is at what level raised BP is a risk which justifies drug treatment. Treatment is usually lifelong as, although the condition can resolve itself, this is unusual. A systolic BP of 160 mm Hg or more and/or a diastolic BP of 95 mm Hg or more warrants at least frequent checks. That raised BP can be well controlled by drugs means this is no reason to reject an applicant or to retire someone early.

A person with raised BP must realise the need for long-term treatment; treatment dramatically improves the outlook.

Screening for hypertension requires standard techniques. A single high reading is significant but must be repeated after a few days. It was once thought that the individual is better not told if the BP is borderline; this was simply noted in the records and a repeat check arranged within months rather than years. The modern trend is to make known the full medical results, both to the GP and to the person who has been screened.

A single reading is a guide to overall BP, though some people's BP goes up when visiting their doctor. No longer is it advocated that if the pressure is raised, the individual should be rested for a quarter of an hour and a further reading then taken when the pressure has usually fallen. The lower reading can give a false impression of normality. Also, no longer is it advised that the individual be told little about the condition and merely handed a prescription. This approach may have contributed to noncompliance in the past. The person felt well and was then given tablets, possibly with side-effects which made him or her feel unwell. Nowadays the drugs used are largely free from serious side-effects.

Once people know they have raised BP they may complain of symptoms for the first time and their sickness absence may increase; but the headache they and most people associate with raised BP is usually the effect of worrying. One survey following *compulsory* screening showed an increase in sickness absence to three times its previous level. This should not be a disincentive to diagnosing the condition by voluntary screening: such surveillance and then treatment can be life-saving.

As stated, usually hypertension is a silent condition until years later, when the resultant stroke or heart attack occurs. Headaches are either unrelated, due for example to referred pain from the neck, or the result of tension and anxiety, especially in those who know they have the condition and worry about it.

It can be appropriate and effective to involve patients actively in day-to-day monitoring of their own treatment. Just as it is standard for diabetics to test their urine regularly for sugar, some doctors now teach patients with raised BP to monitor their own BP, and vary the treatment dosage accordingly.

Heart attack

A heart attack, coronary thrombosis (CT) or myocardial infarction, is a common illness in this country, especially in middle-aged men. Coronary heart disease is responsible for 25 per cent of all deaths in the United Kingdom and kills at least 180 000 people each year. A statistic, like that for lung cancer, in which the UK in general, and Scotland, Wales and Northern Ireland in particular, leads the world.

The British Heart Foundation gets this stark message across by means of a neat calculator. The employer enters the total number of male and female staff and reads off working days lost per year as a result of heart disease, and the direct cost of this to the company in lost production. Figures come from the DHSS and also give the number of potential premature deaths from heart and circulatory disease each year in those over thirty-five. For example, 1000 staff means a loss of 4270 male days and 1470 female days, and the potential deaths of 3.85 men and 0.80 women, at a cost for men of £162 602 and for women of £36 897. 21 per cent of all male absence from work is because of heart and circulatory disease, which is responsible for 141 590 hospital admissions a year of men aged from thirty-five to sixty-four in England alone, and causes 45 per cent of the deaths in men in this age group.

There are a number of risk factors: the more that apply, the more likely a heart attack. They include:

(1) Family history of heart attack before 60.
(2) Cigarette-smoking.
(3) Raised blood pressure.
(4) Raised blood fats.
(5) Lack of regular, energetic, physical exercise.
(6) Stress.
(7) Obesity.

(8) Male sex and middle age.
(9) Diabetes.

Some of these factors are more definitely implicated than others. In the light of evidence from the British Regional Heart Study, Professor Gerald Shaper has devised a simple formula for degree of risk which focuses on smoking, blood pressure, family history, diabetes and symptoms of angina and breathlessness.

There is no particular predisposition of office workers to a heart attack, but a desk-bound pressured job, driving to and from work and watching TV all evening, makes for a stressful sedentary existence; add in a three course lunch and a cooked evening meal, and high blood fats and obesity are almost inevitable. The stage is set.

Lessons can be learned from the following hypothetical case history. A thirty-eight year-old male manager is awakened in the early hours of a December morning by a tight gripping pressure pain in the lower central chest. He had noticed twinges of this pain on going to bed the night before, and three months earlier had felt similar twinges for a day or two, which he had put down to indigestion. This time the pain got steadily worse and became so bad that he telephoned his GP. The partner on call visited at once and prescribed a pain-killing tablet. He was promptly sick, but the pain was now easing away.

Next morning he consulted his GP, who thought the pain likely to be muscular, but since the patient smoked, arranged for an Electro-Cardiograph (ECG) at the local hospital the next day. By then he felt quite well. However, the ECG showed a recent heart attack. He was taken by wheelchair to the coronary care unit (CCU) for two days, then moved to a general medical ward for progressive mobilisation. He left hospital ten days later to continue rehabilitation at home.

This man had smoked forty cigarettes a day; he played desperate squash or badminton twice a week; he had been a top-class sprinter until the age of twenty-one. His father had a heart attack when aged fifty-eight. He owned up to being a tense person who 'bottled things up'. For the previous six months he had worked longer hours than usual, with increased pressure from staff shortages.

His blood fats were normal, so he did not change diet, not even to give up fried food. But he did stop smoking, sucked sweets instead and put on weight. No one gave him much advice about exercise or how to relax. For a few weeks he sat at home fuming from frustration.

After just three days back at work and under pressure again, he noticed chest pain on the journey home; he telephoned his GP and then, as the pain soon cleared, cancelled the appointment. Early next morning the pain was there again. The GP was called out and sent him back to hospital. A second heart attack was diagnosed and the same programme followed as before. Angiogram X-rays of the heart's blood vessels showed partial blockage of two coronary arteries, but good 'collateral' circulation in other blood vessels supplying the heart muscle.

Now he has drastically reduced the animal fats in his diet: food previously fried is grilled, eggs are restricted to three a week, special margarine substituted for butter, and no cream, little cheese and only skimmed milk are taken. He intends to lose the weight he had gained.

He now feels well again, avoids heavy lifting, walks three miles a day without getting exhausted or short of breath, and takes a drug which reduces the work of the resting heart. His self confidence has been severely dented and at present he twitches at the least twinge. He still wants to pursue his career, but hopes to avoid such long hours. He no longer takes work home. In a way he was lucky in that he survived – sometimes without warning there is sudden death.

Although the heart attack is sudden and happens in middle age, the 'silting-up' of the coronary arteries starts much earlier, in childhood even. Over the years, it gradually progresses, with changes in the arterial walls and an increase in the size of the plaques which contain fatty substances, including cholesterol, in arteries throughout the body, not only but particularly in one or more of the three coronary arteries which surround the heart and supply blood and oxygen, to the heart muscle itself. The plaques known as atheroma become obstructive when an area of narrowing restricts blood flow. This may cause symptoms like angina chest pain on exercise, but is commonly symptomless unless and until a clot forms at this point, blocking that artery completely and depriving the heart muscle beyond of blood. This is the sudden heart attack when, accompanied by the sense of impending doom, there is crushing central chest pain which spreads to either or both shoulders (typically the left) and arms, or to the front of neck, or to the back, or the stomach. When severe, there will be clinical shock. There is no such thing as a slight heart attack, as in any such attack there is a risk of sudden death: the heart rhythm may suddenly change or the regular heartbeat and effective circulation suddenly stop. The first few minutes and hours are the most critical. Forty per cent of heart-attack victims die (half have had no warning whatsoever) and forty per cent of those die in the first hour. This is when trained first-aiders can save life, giving the 'kiss of life' and heart massage, which takes over from the ineffective heart and pumps blood round the body, especially to the brain. These manoeuvres sustain life until the experts, an ambulance crew or coronary-care team, take over with specialised equipment. The patient is intubated and ventilated, and drugs and direct DC electric shocks to the chest wall restore the normal heartbeat. In a few areas (like the City of London) there is a coronary ambulance team on call which comes quickly to the scene, gives emergency treatment and then takes the patient to hospital.

Rehabilitation after a heart attack is very different from twenty years ago, when at least three-weeks' strict bed rest was usual. Nowadays, if there are no setbacks, after twenty-four to forty-eight hours on a heart monitor, the patient starts to get up and steadily does more until discharge, in a further week to ten days.

Convalescence and rehabilitation continue at home, with medical guidance about the rate of increase of daily activities, amount of walking and so on. Each schedule is tailor-made. If rehabilitation goes on and on, restlessness, frustration and despondency are likely; if not long enough, there may be a further breakdown in health and restoration of self-confidence is further delayed.

During the recovery period, alongside any medication prescribed, all aspects of life should be looked at critically, searching for clues as to why the heart attack happened. Changes are made in avoidable risk factors to try and reduce the chances of a further heart attack. It would, for example, be foolhardy to smoke at all. The obese must slim to a realistic weight. The heart has been scarred, so it is dicing with death to reduce the oxygen supply by smoking, or to make the heart work harder than it need because of surplus weight. Blood fats are analysed: raised cholesterol, triglycerides and low-density lipo-proteins should fall if the right diet is strictly followed. Whilst dietary changes now are late in the day and akin to 'shutting the stable door after the horse has bolted,' there is evidence that atheroma can be reversed. So a low-fat diet is worth following. Fat spreads, that is, butter substitute, should be polyunsaturated (a reference to the chemical formula), vegetable fats rather than hard, saturated animal fats. One constituent of blood fats – high density lipoproteins (HDL) – is in contrast beneficial. If the cholesterol remains high despite the right diet, then a drug is prescribed to lower it. There is evidence that fish oils are beneficial – the Japanese, who eat a lot of fish, have few heart attacks.

Remember Nathan Pritikin, proved in his late thirties to have advanced coronary artery disease, who devised a strict low-fat, low-sugar, high fibre diet and vigorous exercise regime, and when he died thirty years later from an unrelated cause, had smooth atheroma free coronary arteries.

Return to work should be possible between two and three months after an uncomplicated heart attack. Once over the illness clinically and quite well again, the person is encouraged to put the heart attack behind him or her and get back to a normal life. As with any muscle injury, the healed scar in the heart persists where once there was heart muscle, and will always show on any future ECG. There is little value in forever having check-ups and ECGs 'just to be on the safe side'. The aim is to return to the previous office based job and for a career to continue. If work is physically demanding, there should be three months on alternative duties with no lifting or carrying of anything heavier than fifteen lbs; but if someone's livelihood depends on lifting and the person is symptom-free, then a calculated risk will have to be taken. Overtime should be avoided for several weeks.

Having had one heart attack, the risk of another is greater. This is an inescapable fact and the possibility must be faced; total reassurance that all will be well cannot and should not be given. If it is, and a second heart attack then occurs, there is both loss of faith by all in the medical profession and, if survived, profound loss of self-confidence.

The excess risk of death for those who survive the heart attack to leave hospital, declines over the next ten years from thirty times in the first year to no increased risk.

After the life-threatening experience of a heart attack, a few symptomless people are reluctant to do almost anything in case it causes another, this time fatal, attack, and, regard themselves as near-invalids. Another unhealthy reaction after a heart attack is a straining determination to prove to the world 'I'm as good as ever – if not better'; to play two rounds of golf rather than one, take on extra work and so on.

The middle line is the right line, gradually and sensibly increasing activities under medical supervision.

Details of work should be discussed with the company doctor and manager, preferably before or immediately on return to work. To be avoided is the situation where any non-urgent work has been allowed to pile up, and there is then the daunting task of working through this backlog. As was said in an earlier chapter, and certainly bears repeating, return to work after any long-term sickness absence should start in the middle of the week, leaving only two or three working days before the weekend. Hours of work for the first two full weeks should be where possible cut from eight to six a day, for instance, be from 10 a.m. to 4.00 p.m., including the normal lunch hour. This allows the city worker to travel outside the commuter rush hours. These measures soften the initial stress of getting back to the work routine which contributes to the common complaint of undue tiredness.

In some jobs such a regime is impossible, as is the instruction only to undertake 'light duties', whatever that vague term means. The person either has to do the job completely or not at all. In this situation a longer convalescence is needed. What must not be allowed to happen is the attempt to cram a full day's work into the shortened working day. The full one-hour lunch break should always be taken, when, besides relaxing over a meal, there will be time for a brisk walk.

In summary, someone who has had an uncomplicated heart attack should be ready to return to work during the third month after the attack. Provided the job does not involve heavy physical work this means full normal duties. As after any serious illness or operation, it is sensible for the return to work to be taken in easy stages:

(1) Start mid-week in the first week, so that after two or three days there is a weekend to relax.
(2) Initially avoid rush-hour commuting.
(3) No overtime for several weeks.
(4) No taking work home.
(5) No lifting heavy weights.

Survivors of heart attacks often say that the twelve to eighteen months beforehand included a number of stressful and/or unpleasant life events. There was increased pressure at work, or domestic worries, or strife, or all of these. Sometimes the normal workload appeared more stressful because the person felt unwell. They also often remember feeling irritable, worn out, fretful and 'down', with lack of interest and sleeplessness; common enough from time to time in most people, and so not enough in their eyes to warrant a visit to their doctor.

One study of heart-attack survivors confirmed these behaviour changes in the preceding year. The person does less and gives up inessential activities previously of great interest, but despite trying to conserve energy, feelings of weariness and tiredness remain.

There are other causes for such behavioural change, but these are clues for individuals and management to look out for.

There is nowadays little dispute about the significance of the relationship between dietary fats and coronary heart disease (CHD). Blood fats (lipids) include cholesterol, triglycerides and lipoproteins split into High Density (HDL), Low Density (LDL) and Very Low Density (VLDL). HDLs protect against formation of atheroma which narrows and so contributes to the ultimate blockage of arteries, including coronary arteries, and causes heart attacks.

A diet rich in animal (saturated) fats raises the blood-fat level as so does emotion and reaction to stress. Those with consistently raised LDL blood-fat levels are at increased risk of a heart attack, but what has not been shown is that lowering the blood fat level increases longevity. The Royal College of Physicians' report on prevention of CHD recommends a reduction in dietary fat from 35 per cent to 30 per cent and a switch from dairy products to polyunsaturated fats. There has been concern that lowering blood fats may actually increase the incidence of certain disorders of the liver and gut, but I am sure we should all eat less fat and more fibre in the form of vegetables, fruit and wholemeal bread.

The incidence of sudden death from a second heart attack in the five years after the first is lower in those who are physically fit. So moderate exercise such as walking is part of the rehabilitation programme after a heart attack. A planned, progressive and, ideally, supervised, exercise programme leads to a greater degree of physical fitness than leaving individuals to their own devices with a sheet of written instructions. When exercise becomes hard work, with sweating or feeling tired, it is time to stop for the day.

Such strenuous exertion must only be gradually incorporated in a training schedule. Sudden unprepared exertion, especially 'isotonic' with little or no movement, such as straining to lift a heavy box, can precipitate altered heart rhythm or even a second heart attack.

The relationship of heart attack to stress is controversial. A research project in America some years ago concluded that the population is divided into two personality types – A and B. Type A are said to be six times more liable to a heart attack than type B. A type-A personality is extremely time- and deadline-conscious, with a chronic sense of time urgency and a wish to dominate; he or she is highly competitive, fringing on the aggressive, and addicted to work –

a workaholic. Type B still works hard and is ambitious but does not have the same need to assert power. Type A should be popular with an employer as, until the heart attack, he or she gets results fast and has an excellent sickness absence record, seemingly an ideal combination. 'I'll do it if it kills me' is a classic type-A statement. Unfortunately, it sometimes does.

The long-term study into the incidence of heart and circulation disorders in Framingham, USA, gives a somewhat less dramatic picture in that, taking other risk factors into account, the incidence of CHD is twice as high in type-A men than in type-B men, and four times as high in type-A women than type-B women. More recently a Type-A high risk subgroup has been identified in which the person is forever hostile, aggressive and angry, alert for a stab in the back. Maybe it is in this group that there is particular vulnerability to a heart attack. A phrase to summarise Type A is 'anger in' and Type B 'anger out'. It must be remembered that a certain amount of stress, challenge or pressure is very necessary, and a totally stress-free life is neither possible nor desirable. However, type-A people should be encouraged to move closer to type-B responses and to learn to relax. This is something many people, who are well able to relax at home with a sport or hobby, cannot do in a work setting – after a difficult phone call, interview or meeting, they spend the rest of the day in a state of tension.

Sometimes after a heart attack there is less than full recovery in that the person gets angina. The diagnosis is suspected from the history, as examination may be normal. Angina pain typically spreads from the centre of the chest to the neck or left shoulder and arm, brought on by exercise or by emotion and stress. At rest, sufficient blood gets through the diseased coronary arteries but on exercise the heart works harder, beating faster, and the blood flow is insufficient for the heart's needs, so heart-muscle pain occurs.

Angina may remain fairly mild, as pain coming with heavy exertion, when hurrying up a steep hill for example, or it may be progressive to the stage where chest pain stops a man in his tracks after slowly walking 100 metres on the flat. An important clue is that such anginal pain clears soon after exercise stops. Of the very many other causes of chest pain, pleurisy, hiatus hernia, shingles and chest-wall muscle strain are four common ones.

People with chest pain from some other, relatively innocent cause, such as muscular strain or referred pain from the back, may worry that the pain is coming from the heart. This pain may be nothing more than stress and muscle tension. Medical treatment for angina can be very effective, but severe angina may delay or prevent return to work. Angina on exercise can be so disabling as to make commuting impracticable. Commuting becomes difficult when there is no alternative to the station steps or a hill to be climbed. Similarly, physical work such as office cleaning may become impossible.

Sometimes in a large company, a job with lighter duties can be created, but in the long-term other staff may resent this, particularly if pay is linked to total productivity and the 'passenger' reduces everyone else's earnings, or the angina victim feels awkward, watching others do the work. Angina qualifies for the disablement register.

Where people are likely to remain medically unfit to do the work for which they were employed, early retirement or termination of contract on health grounds will need to be considered, if suitable alternative sedentary work close to home cannot be found. This also applies where the angina is brought on by work stress. There are no stress-free jobs in modern industry.

In some cases, where medical treatment is ineffective or angina seriously interferes with normal life, coronary angiograms (X-rays) pinpoint the sites of atheromatous narrowing in the coronary arteries. Using open-heart surgery, leg veins and/or the internal mammary artery are grafted on to the heart wall and into the circulation to bypass these blocks. Blood supply to the heart muscle is improved and angina ceases. The patient once again has a pain-free active life and will be fit to commute and work normally again. If just one vessel is narrowed, angioplasty squashes the atheroma and opens up the artery again without the need for an operation.

The launch of the DHSS/Health Education Authority 'Look After Your Heart Campaign' in 1987 should, in the coming years, reduce the incidence of heart disease in the United Kingdom.

Stroke

Each year about 100 000 people in Britain have a stroke and about 50 per cent of them die within a year. Of the survivors, a third remain bed-bound, a third recover completely and a third recover partially.

A stroke – brain damage caused by disturbance of the blood supply to the brain – is the commonest cause of chronic physical and mental handicap in the western world, particularly among older age groups, but also in wage-earners. Between a quarter and a third of stroke victims are under sixty-five.

The many types of stroke are collectively called cerebrovascular accident (CVA); the commonest is due to a leak of blood from, or clotting of blood within, an artery supplying a part of the brain. Common to the many and varied effects of a stroke is the fact that brain cells starved of blood for more than a few minutes stop functioning permanently. The result is, commonly, weakness or total paralysis of one or both limbs on the opposite side of the body to the area of brain affected; sometimes the face is involved also. If the dominant side of the brain is affected – the left for a right-handed person – speech is usually also affected.

An extensive stroke is likely to cause disturbance of memory or intellect. In half the cases, raised BP is a contributory factor.

The basis of treatment is rehabilitation and prevention of complications and recurrence. Skilled and immediate physiotherapy, encouragement and restoration of morale are vital. The aim is for as normal a life as possible.

The most rapid improvement usually takes place within the first ten days. A prognosis about return to work should not be expected for many weeks.

There are special stroke units and centres for speech therapy.

MUSCLE AND JOINT DISORDERS

Arthritis

Rheumatism is a vague term which means very little other than any pain or discomfort from a joint, ligament or muscle, not caused by injury. Certainly it has nothing to do with rheumatic fever or rheumatoid arthritis with which the term rheumatism is confusingly similar. Pain in the back is often called rheumatism as is osteoarthritis as well as rheumatoid arthritis.

Rheumatoid arthritis can cripple but more usually waxes and wanes. In time, rheumatoid arthritis may burn itself out. The cause is unknown. Women are affected more than men. Typically rheumatoid arthritis starts gradually in early middle age in both hands and feet with symmetrical involvement of small joints. Other parts of the body may be affected. Much joint deformity can occur, but often useful function is preserved. Medical and sometimes surgical treatment helps, but drugs can cause side effects. The hallmark of rheumatoid arthritis is morning stiffness, so it is hard to get going, and manual work or work involving standing or using the hands skilfully, such as typing, may be difficult. A temporary late start time when the disease is particularly active may allow someone with rheumatoid arthritis to keep their job.

Osteoarthritis should be called osteoarthrosis as there is no inflammation and '-itis' is inappropriate. Osteoarthritis can involve any joint, often one or both knees and/or hips. Also commonly affected are the spine and finger joints. The condition typically presents in middle and old age. Pain from osteoarthritis in the legs is more noticeable on exercise or on first moving after a rest. Osteoarthritis results from wear and tear changes or from earlier injury to joint surfaces and cartilage. Sometimes one knee or hip only is affected, although the other has had the same wear and tear. Osteoarthritis is also gloomily called degenerative joint disease. It cannot be cured. If drugs and physiotherapy do not help enough, surgery may, such as replacing the worn joint with an artificial one. An operation is not always the answer, so a decision about unfitness to do a particular job or to commute because of arthritis should not be influenced by the possibility of future joint replacement surgery.

Gout

Many more men get gout than women, in the ratio of fifteen to one. There is an acute reaction to sharp crystals of uric acid with sudden pain in one joint – classically the base of the big toe – but also the ankle, knee, wrist, elbow or a finger joint. The joint is exquisitely tender for days, but drug treatment is rapidly effective. Frequent gout is prevented by another drug which lowers the level of uric acid in the blood; diet has little effect.

Back pain

Back pain can be a twinge of no consequence, an acute severe disability which comes and goes, or grinds on and on, becoming chronic.

Musculo-skeletal disorders, of muscle, tendon, ligament, bone and joint, make up 15 per cent of sickness absence; back pain accounts for 60 per cent of this, second only to bronchitis as the commonest reason for absence from work. On any day, 50 000 people are off work with back pain. The cost is enormous. It might reasonably be thought that the incidence of occupational back pain is highest where the work involves lifting – in mining, dockyards, the construction and building industry, but these workers are more likely to be physically fit and to know how to lift properly, so it is rather sedentary workers who are called on to lift from time to time who are, if anything, more likely to injure their backs. Almost half of the job-related injuries in nurses involve the back; in up to 80 per cent of these, lifting of patients is implicated.

Back disorders cause sickness absence in any company. The sedentary office worker who bends or lifts awkwardly may be incapacitated for weeks. The employee may claim back pain was brought on by a particular activity, incident or accident at work, but of course back pain is not only work-related; the back can be strained anywhere, anytime, and home DIY or sporting activities are sometimes the cause.

Terms used for back pain include rheumatism, lumbago, sciatica, slipped disc, disc lesion and prolapsed intervertebral disc (PID). A recurrent disc disorder often starts in young adults; it causes episodes of pain in the back which may radiate down one leg. A disc disorder in the neck causes pain which may radiate down one arm. Many of these episodes clear spontaneously sooner or later; treatment meantime relieves symptoms and can prevent recurrence by improving posture, as may the teaching of remedial exercises.

The concept of modern physiotherapy is not only to treat symptoms, but also to look for the cause and where appropriate to give preventive, educational advice. The leaflet for staff recovering from acute backache, reproduced at the end of this section, is an example of such advice.

The bones of the spine do not form a straight line, but a reverse S curve evolved from one gentle concave curve at birth. The neck bones curve forward from when a baby holds its head up unaided, and the lower back bones also curve forward at the stage of sitting up and standing.

The commonest sites of problems are the lower back and the neck. The lower back is made up of five lumbar vertebral bones, the cushioning discs between each, plus the disc between the fifth vertebra and the sacrum (a triangle of fused bones at the end of which is the coccyx, the vestigial tail).

This lumbar part of the spine curving forwards creates a hollow in the small of the back. The shock-absorber discs have a fibrous outer layer and a pulpy core. Damage to a disc causes many of the serious and recurrent back problems.

Often there has been a gradual, symptomless weakening of the disc, and what finally causes the 'slipped-disc' syndrome is some minor 'last-straw' action, such as bending over in the garden or to tie a shoelace.

A tear through the fibrous outer layers of the disc allows the core to extrude and press on one or more of the nerve roots which are close by. These nerves, supplying power and feeling to the legs and pelvic contents, emerge from the spinal cord, which is enclosed and protected by the spinal bones. Most commonly affected is the sciatic nerve. Pressure on this nerve produces 'referred' pain down the back of the leg and into the foot.

There is sudden onset of back pain and the spine 'locks', causing difficulty in straightening up. The pain may remain in the back or move to one leg; typically the pain is made worse by a cough or sneeze, as well as by bending.

In some cases, there will have been an unsuspected, undiagnosed, underlying weakness of the spine, either congenital (from birth), or acquired in early life. Examples are: where one vertebra slips forward over another (spondylolisthesis), and increased curvature of the spine (lordosis); to be overweight or to wear high-heeled shoes exaggerates lordosis. Occasionally, and this is a possibility always to be excluded before treatment, especially manipulative physiotherapy, back pain is due to cancer in a vertebral bone, leading to bone pain or nerve pressure or compression of the weakened vertebra.

A less serious, but no less painful, cause of back pain is injury to back muscles or ligaments; this pain remains localised to the site of injury. A direct blow, or landing from a height on the heels, may also damage one or several vertebrae.

The spine is supported and strengthened by muscles in the back, abdomen and buttocks. If these muscles weaken through lack of use, or are overstretched or fatigued through overuse or misuse, their protective effect is lessened and more strain is placed on the spine.

Initial, orthodox, conservative treatment involves bed rest and pain-relieving drugs.

The traditional treatment for a disc disorder is to lie day and night on a hard, supporting surface, waiting for the pulpy core of the disc to return to its normal position. The outer disc has no blood supply and cannot heal or scar, so the spine is never thereafter as strong as it was.

A surgical corset supports, but is cumbersome, can be uncomfortable, and allows the muscles under it to become flabby; its main value is to remind the wearer to keep a good posture and not bend from the waist.

Infrared heat from a lamp soothes and temporarily eases muscle spasm and pain, but gives no lasting benefit and is really no better than a hot-water bottle or the warmth from a nearby fire. Physiotherapy which includes manipulation and traction can be effective sometimes. 'In-patient' treatment may be necessary for total bed rest – this is seldom possible at home – and continuous traction. Other treatments include 'epidural' injection of a local anaesthetic into the spinal canal, and injection of a sclerosant which forms scar tissue to strengthen the back.

Back strengthening is also achieved by teaching exercises to be done daily. Opinions differ about which exercises are best. Bending over to touch the toes can strain the back. Whilst strengthening supporting muscles helps, it is most important to use the back correctly. The earlier this is learned the better and the more likely that back strain is avoided. Having suffered a serious attack of back pain, it is advisable to avoid activities at home and at work which put any strain on the back, lifting heavy suitcases or furniture, for example.

The overweight should lose weight to the normal for their height. Being two stones overweight is equivalent to lugging around forever two heavy suitcases.

If nerve pressure from a disc lesion affects bowel or bladder function or causes foot drop (inability to lift the foot), or unrelieved pain, an operation – decompression or removal of the disc (laminectomy) – is needed. After an episode of back pain, it is largely the patient who decides when he or she is able to return to work. Psychological factors can be relevant.

Fictitious backache is unusual, but there can be an exaggeration of the symptoms, especially when someone with a history of proven back disorder is under stress. Some people are genuinely susceptible to recurrent attacks of incapacitating back pain from a spinal disorder, others have a low threshold for pain, or their complaint of back- or neck-ache masks an underlying depressive illness or anxiety state. It can be difficult, if not impossible, for the examining doctor to accurately gauge the severity of the symptoms, particularly if there are X-ray 'wear and tear' changes. These X-ray changes are permanent and, in the neck particularly, are common in anyone over forty, most of whom thankfully remain symptomless. Of course, normal X-rays do not necessarily prove there is nothing wrong.

As previously stated, after one bad attack of backache it can happen again. Where possible, a change in duties to avoid lifting heavy weights is a sensible precaution. Teaching the correct ways to stand, lift and carry will go far to prevent recurrence.

On return from sick leave for backache, the individual may have been advised to avoid 'heavy lifting' and may have been given a note to this effect. Unless defined, this phrase means little – what is heavy to one person is not heavy to another; much better is to specify an upper limit in the range of, say, 7 to 10 lb.

Prevention involves instructing new staff in the correct techniques of lifting and handling, with periodic reminders, and promoting working conditions which minimise back strain. Trade unions help by emphasising the importance of following procedures which reduce the risk of back pain. The need for adequate instruction is covered by HASWA.

The Post Office limits the maximum load lifted by one man to 48 lb (22 kg). Guidelines should not be limited to manual workers; sedentary office workers are also at risk – picking up a heavy ledger from a low shelf or moving a pack of stationery or a typewriter, are examples. Poor posture alone will, in time, strain the back.

General advice is:

(1) Stand and walk 'tall'.
(2) Head up, chin in.
(3) When moving anything heavy use mechanical aids, such as a trolley, wherever possible. If this is impracticable:
    (a) use the whole body;
    (b) face the direction you will be going in;
    (c) feet apart;
    (d) load close to the body;
    (e) keep the back straight, not necessarily vertical.
    (f) bend one or both knees, and not from the waist, so that the leg muscles do the work;
    (g) take equal care putting the load down.

A practical demonstration is the best teaching method, but simple written advice will help. Posters and a personal reminder card reinforce the advice given. In some companies, a leaflet sent out from the furniture store with every new chair supplied is also issued to every new entrant, to fulfil a requirement of HASWA to instruct those required to lift at work (and that includes just about everyone) on the correct methods of lifting and handling.

Preventive information in the form of a leaflet, poster and training manual is available from the Back Pain Association, Grundy House, Somerset Road, Teddington, Middlesex, TW11 8TD. This association is a charity set up in 1969 by a businessman, Mr S.W. Grundy, with the objectives of encouraging research into the causes and treatment of back pain through the Society for Back Pain Research; raising money for this work; teaching the right ways to use the back, and forming local groups to help those liable to back pain.

As yet there is no specific legislation for office staff as to what weight they should reasonably be expected to lift when in good health. There can only be guidelines, not rigidly applied limits: much depends on the individual's physique. After the passing of the Sex Equality Legislation, men are it seems generally less inclined to help women carry loads: 'if they are equal to us they can lift it themselves'.

The OSRP Act, Section 23(44), like the Factories Act, Section 72 (1), gives sensible advice on limits: 'No person may be required in the course of his work to lift, carry or move a load so heavy as to be likely to cause injury'.

The 1982 proposals for Health and Safety (Normal Handling of Loads) Regulation and Guidance give the following specific guidelines:

Below 16 kg (35 lb)
No special action required, provided those relatively few individuals likely to face serious risks when handling weights of this order have been identified.
From 16 kg (35 lb) to 34 kg (75 lb)
Administrative procedures required to identify those individuals unable to handle such weights regularly without unacceptable risk, unless mechanical assistance is provided.
Above 34 kg (75 lb) to 55 k (120 lb)
Unless the regular handling of weights of these magnitudes is limited to effectively supervised selected and trained individuals, mechanical handling systems should be employed.
Above 55 kg (120 lb)
Mechanical handling systems should always be considered at this level. Where not reasonably practicable, selective recruitment and special training is essential, since even with effective supervision, very few people can regularly handle weights of this order with safety.

Pain in the mid-back is unusual, but at any age symptoms from the neck – the cervical spine – are very common. Pain may be just in the neck or also in one or both shoulders or arms; headaches may occur. Lay diagnosis is that this has resulted from sitting in a draught, but whilst a draught aggravates muscle spasm, it is not the true cause. Lifting awkwardly can injure the neck as well as the lower back. The medical diagnosis may be 'cervical disc', but from middle age onwards, the joints of the neck bones show 'wear and tear' changes, called cervical spondylosis, and whilst this is often symptomless, there can be chronic or recurrent pain in the neck, or in the head, shoulders or arms because of nerve-root pressure which can also cause weakness or changes of feeling in one or both hands and arms. The true cause of such symptoms can be difficult to distinguish from localised arm disorders like carpal tunnel syndrome or tenosynovitis. A cervical disc gets better on its own or with physiotherapy, but in cervical spondylosis, if X-rays show the typical changes which are permanent, there is no cure, though drugs and physiotherapy can relieve symptoms. Though helpful for acute neck pain, it is doubtful whether the long-term wearing of a polythene or foam collar to support the neck and restrict movement, is of benefit. Surgery is rarely carried out, partly because of the close proximity of vital nerves, arteries and veins. Prevention of recurrence of a neck disorder is more problematic than prevention of low back pain, as the neck is moved and twisted so often in everyday life. Someone liable to recurrent disc disorder or symptoms from cervical spondylosis should try, as far as possible, to avoid undue neck movement, and not to carry anything weighing over 10 to 15 lb. Any load should be evenly spread between the two arms and held close to the body.

At the work station, it is essential to sit properly, and if at a keyboard and screen, to have the work well located to avoid undue bending of the neck. A document-holder should be used and the screen angled so that the user looks down about 20 degrees from the horizontal to the centre of the screen. Paper work on a desk can be angled up on a lectern plinth to avoid bending the neck.

First aid for neck pain includes wrapping the neck with a rolled towel for support (especially useful at night); experimenting with the number of pillows, and, if one is used, a ribbon tied round the centre gives additional support to the nape of the neck. A range of drugs can reduce the pain. As with low back pain, a heat lamp (usually infrared) or ice pack eases muscle spasm but there is no permanent benefit. A hot-water bottle may be as good.

Backache is common in professional drivers. Driving for a living is not a good idea for someone with a history of a serious back disorder, either in the neck, as traffic surveillance and reversing require a full range of neck movements, or in the lower back, as the driving position may aggravate backache there. Bending and twisting whilst loading and unloading heavy, awkward packages in and out of a car boot or off the back seat or shelf can strain the lower back and the neck. In the past, the bucket seat in a car was no support to the back, but now most cars have seats giving good low back support, providing of course the driver correctly adjusts the seat for his or her frame and height. The same need applies to the office chair; most chairs are adjustable if the person using the chair takes the trouble to find and use the controls.

To repeat, anyone with recurrent back or neck pain should not do work which involves repeated bending or twisting, or the lifting or carrying of loads of more than 10 to 15 lb.

Pre-employment medical examination of applicants for jobs which involve using the back, for instance porters, should reveal significant past or present back disorders. Obviously such work is far from ideal for a person who has such a history.

In a claim for compensation for industrial injury from back injury caused by work, a past history of backache is significant and could affect the liability of the employer. In all such cases, enquiry by the company doctor, with consent, of the GP is indicated. A person with spinal canal stenosis is at particular risk. The emergency aid guidance notes on pages 258–9 were compiled by Miss Susan Barnes, a sessional physiotherapist at Nat West's Occupational Health Centre.

CHEST DISORDERS

Chest or respiratory disorders account for a third of consultations with GPs and 20 per cent of their prescriptions. At any time of the year one person in five has a respiratory disorder; most will have an upper-respiratory-tract infection (URTI) – like a cold; serious illnesses such as pneumonia and lung cancer make up a small percentage of the total, but feature high on the list of causes of death.

URTI covers such conditions as tonsillitis and laryngitis, sinusitis and coryza, the common cold. An URTI is usually due to one of a number of viruses and is the commonest illness in Britain.

Influenza vaccination neither protects against, nor induces, an URTI. URTIs occur most frequently in children starting school;

ACUTE LOW BACK PAIN – Why it happens – how to survive an attack – and how to stop it happening again.

TO BE READ IN CONJUNCTION WITH ADVICE AND EXAMINA- TION BY YOUR GENERAL PRACTITIONER (WHO SHOULD ALWAYS BE CONSULTED).

At some time in their lives, almost everyone will have suffered from a strained back – hardly surprising considering the complexity of the spine and the strain put on it in daily life. There are many causes of back pain but in practice, back pain is usually due to a mechanical problem in the spine after a sudden strain such as lifting awkwardly – sometimes the 'last straw' to an already insidiously damaged spine from an accumulation of minor injuries. Poor working positions, bad posture and being overweight contribute.

The pain may radiate into others parts of the back and into one or both legs. Sometimes pins and needles and numbness occur.

Spasm of the back muscles often accompanies acute back pain.

The sequence of pain causing spasm causing more pain has to be interrupted to relieve the attack.

In the early, acute stage of an attack of low back pain, any movement is agony – so lie down on the floor or on a bed with a firm base and good supporting mattress, or if at first you cannot tolerate a bed, lie on a softer settee. You may find the only comfortable position is on your back, knees bent, knees supported by 3 pillows. Try to maintain the natural hollow curve above the base of the spine.

These other lying positions should be tried as soon as possible to help preserve this curve:

a) On your side with 2 pillows for your head and 1 between your knees.
b) On your stomach with 1 pillow under your tummy.

For acute back pain, cold is better than heat; put an ice pack or packet of frozen peas in the hollow of your back for 5 to 15 minutes, depending on coldness (put a tissue between your back and the cold pack). After a few days, or if cold treatment is ineffective, use heat from a hot water bottle for a similar time.

Move as little as possible and only get up to go to the toilet.

Getting out of Bed

Don't sit up – roll on to one side; gently lower legs over the edge, supporting yourself on the lower arm and elbow; push down with the upper arm to sit; come to the edge of the bed and push down with your hands to help you stand.

Sitting

This increases the pressure on the discs in your back so only sit for a short time in a firm, upright chair. If you sag or slouch, this will increase back pain.

Put a firm cushion, rolled up towel or special lumbar support in the small of the back. Get up and move around every 10 to 20 minutes so that you don't get 'set' in one position. Do this by lifting your bottom to the edge of the chair and using your hands to push down on the chair, stand up gently. Occasionally, someone suffering from acute low-back pain has to crawl on all fours.

Coughing or Sneezing.

If about to cough or sneeze, lean backwards to hollow the small of your back.

Lifting Anything

DON'T!

Bathing

Getting in and out of the bath is very difficult – have a shower or a wash down instead until the acute pain has eased.

Bending Forwards

Don't; if you have to get to a cupboard or pull up bedclothes, kneel and keep the back arched.

Toilet

Sit upright on the toilet – if necessary, brace your hands on the wall. Don't get constipated – straining increases the pressure in your abdomen and on your back. Eat plenty of fruit and other high fibre foods like vegetables and wholemeal bread, and bran cereals for breakfast.

Reminders and Guidelines

Take infinite care of your back, and you have every chance of a speedy recovery.

Don't try to 'work off' your pain, it will only get worse.

After the acute pain has eased, resume normal activities gradually.

Try to keep the normal hollow in your lower back with an erect posture.

Keep your back straight and supported when sitting – you may need a back support – a doctor or physiotherapist will advise.

Don't drive during the acute stage.

A gentle exercise routine can be implemented under the guidance of a physiotherapist who can also advise on the value of a corset for activities like gardening.

© Occupational Health Department, National Westminster Bank Group, 1987.

eight infections a year are not unusual. Older children and adults have on average two or three URTIs a year.

Respiratory disorders account for 25 per cent of absences from work because of illness, excluding short absences of three days or less, but including invalidity – this explains why bronchitis makes up 50 per cent of the total; URTIs contribute 30 per cent. With the more serious respiratory conditions, the decision whether or not to stay off work is easily made. The person feels ill, wants to stay in bed and is content to follow 'doctor's orders'. With URTI and throat infections the decision is more difficult and there is debate about whether to stay at home or to go to work and pass on the cold to others. Germs are transmitted by coughs, sneezing, even touch, so by coming to work the sick person puts others in the same train compartment and office at risk; anyway efficiency at work may be impaired. Although there is no proof that it does anyone harm being out in all weathers, a commuting journey will not help a cold get better; on the other hand, there may be no one at home to look after the sufferer, and being alone at home may be depressing and in winter, cold and expensive in heating bills. It is mostly physical contact, such as shaking hands and then touching the nose, where the cold virus thrives, which spreads colds. Most career workers work with a cold. Treatment for a cold means treating the symptoms, that is, by warmth, inhalations, rest, plenty of fluids and something to suck.

Tonsillitis

An attack of acute tonsillitis is either due to a virus or to bacteria; it needs a throat swab to sort out which. Bacterial infection needs an antibiotic whereas viral infection does not.

Recurrent tonsillitis may cause repeated sickness absence for a few days each time from school and later, from work. An ENT surgeon may advise removal of the tonsils, but less readily nowadays than in the past.

Bronchitis

Acute bronchitis affects the larger air passages in the lower respiratory tract, and after several attacks can progress to chronic bronchitis, which is classically linked to smoking and to air pollution.

Chronic bronchitis is looked on as the British disease, but this is a false impression, as criteria for diagnosis vary in other countries and respiratory conditions make up the largest group of disorders all over the world. Features of chronic bronchitis include shortness of breath, cough, and sputum. These are worse in the winter and last for at least three months of the year for at least two years.

Tuberculosis

Tuberculosis of the lung (TB) is nowadays less common in the United Kingdom, but still occurs; if it does; there is the need to check whether the disease has been passed on to anyone else at home or at work. For this to happen, TB must be 'open', that is, with tuberculosis germs seen in or grown from sputum samples, in which case the Environmental Health Officer (EHO) is notified and arrangements made for chest X-ray and skin testing of close contacts. It can be difficult to know where to 'draw the line'. Usually only those who have regularly been in the same office as the TB patient need to be X-rayed. There need be no hurry to do this, and even benefit from some delay, as it takes some time for the changes to become apparent on X-ray. In countries where AIDS is prevalent, TB is rampant.

Asthma

Asthma is unpredictable. Acute attacks of airway spasm, especially at night, cause tightness in the chest with wheezing, coughing and breathlessness. Each attack usually clears quickly, either spontaneously or with treatment – but each year 1500 die from asthma, usually suddenly and unexpectedly. A not uncommon trap is overdose of a drug which has become ineffective. In some kinds of asthma there is no outside cause; in others many factors can trigger an attack. What one person is susceptible to may not affect another. Trigger factors include a chest infection, chemical irritants, exercise, coughing, laughing, cold, dust or smoke. Asthma is an occupational hazard in a number of industries. In a quarter of cases there is a major allergic component, and in 40 per cent emotion plays a part, aggravating rather than causing an attack. Asthma may be prevented by taking prophylactic medicines; prompt medical attention is needed should there be a severe attack.

A prospective employee may disclose a history of asthma in childhood, along with eczema and hay fever which may run in the family.

This sort of asthma is likely to clear completely in the teens but may recur many years later. The outlook for late onset 'intrinsic' asthma is less good, particularly when there is a history of chest infections.

Hay fever

In the United Kingdom, hay fever affects as many as three million people each year during the 'season', which normally lasts from the first week in June until the third week of July. The title is misleading, as hay fever is neither caused by hay nor produces a fever. Rather it is an allergic reaction to one or more pollens which act on sensitive cells in the nose resulting in the familiar sneezing, streaming nose and red, watering eyes. Cool weather in April and May shortens the season. Less than a quarter of sufferers seek help from their doctors, possibly mistaking hay fever for a summer cold.

A course of twelve desensitising injections given each year every ten to fourteen days, well before the hay-fever season starts, may help, but there is significant risk of a serious reaction; deaths have occurred. Because of this risk, the person who has had such an injection is required to remain under observation for three hours, with full resuscitation equipment and expertise available in case of collapse. Antihistamine drugs improve all the hay-fever symptoms, but some can cause drowsiness and so affect concentration; more recently drugs without this side effect have been marketed.

Influenza

Influenza is often wrongly given as the self certification reason for a few days' sickness absence. Influenza usually occurs in winter in outbreaks. It is a most unpleasant, characteristic illness which starts suddenly with shivering, 'feeling rotten', then fever, headaches and painful arms, legs and back, a flushed face and sore 'heavy' eyes. A cough and sore throat are usual. The appetite is lost.

With bed rest (and the person with flu needs no persuading) and a pain-relieving drug such as aspirin or paracetamol, the patient starts to pick up in a few days and is away from work in all for about a fortnight.

INFECTIOUS DISORDERS

AIDS

A great deal of world attention has been focused on AIDS with the number of cases doubling every ten months. By mid 1987, in the USA 36 000 people had developed the disease, about half of whom have died and more than 1½ million are estimated to be carrying HIV, the AIDS virus. In the United Kingdom the figures were around 1000 and 100 000 respectively.

In the future, America and Europe may be faced with many more deaths; some African countries may suffer severe depopulation. Worldwide, in 1987, AIDS cases were estimated as high as 150 000, with up to 10 million people infected with the virus.

In that year an article in National Westminster Bank's 'in-house' newspaper, *Bankground*, gave the following facts because myths and misunderstandings about AIDS and how it is spread abound. The following notes answer commonly-asked questions and clarify uncertainties.

What is AIDS? AIDS stands for Acquired Immune Deficiency Syndrome. *Acquired* means caught from someone else rather than inherited; *immune deficiency* indicates that the body has partly lost the means of defending itself with antibodies against certain illnesses; *syndrome* is the pattern of illnesses which result. These are mostly rare infections and include a serious lung condition, pneumocystis carinii pneumonia, and also a rare cancer which affects the skin and other parts of the body – Kaposi's sarcoma.

What causes AIDS? AIDS is caused by the Human Immunodeficiency Virus (HIV), which like any other virus lives within body cells. HIV gets into certain white blood cells which provide immunity (particularly T-helper lymphocyte cells), incorporates itself into the nucleus and waits there dormant for what can be many years, until certain other virus infections stimulate the cell. This triggers HIV to multiply and break out of the cell, leaving sieve-like holes so that the cell dies and immunity wanes.

When did AIDS first appear? In 1981 the first few cases of AIDS were diagnosed in young homosexual men in the USA. Two years later HIV was discovered.

How can a person catch AIDS? It appears that HIV is only passed on by semen, vaginal secretions, and blood. Most cases of AIDS in the UK so far (over 90 per cent), have been in homosexuals, but hetero-sexual intercourse with an infected woman or man can transmit the infection. A mother who is HIV-positive can pass the virus on to her baby during childbirth; 50 per cent of such children develop AIDS within two years, and the mother is herself more likely to develop AIDS.

Blood-to-blood transfer principally involves intravenous-drug-abusers sharing needles and syringes, but there are haemophiliacs and others infected by blood products imported from the USA for transfusion before this method of transmission was realised.

The risk of catching AIDS from giving first aid is negligible, and in world medical literature there is no such case described. Any risk that there might be would arise from failure to observe good personal hygiene. First aid leaflets should remind first-aiders to cover any cut or abrasion on their own hands with a plaster, to wash their hands before as well as after treating a minor injury, to wear gloves when cleaning up spilled blood and immediately to wash off any blood splash on their skin.

How can a person tell if he/she has caught the AIDS virus? Infection with HIV may cause an initial glandular fever-like illness. About three months later an antibody response occurs which does not destroy HIV but can be detected by a blood test. A negative test proves nothing and for many people – 'the worried well' – is un-necessary.

Is there any risk of catching AIDS from working with a person who has AIDS or is HIV positive? AIDS is not passed on by normal social contact, by shaking hands, or by a cough or sneeze. There is no risk from sharing cups, glasses, or towels, from using the same lavatory or wash-hand basin, nor from eating food prepared by such a person. There is no risk either from handling such a person's papers or money.

Is there any other risk of catching AIDS? There is no risk from a swimming pool and no risk from an insect bite. If you undergo any procedure which may draw blood, the equipment used should have been properly sterilised. This includes dental treatment, chiropody, acupuncture, ear-piercing and tattooing. In the home, the only risk would be from sharing a razor or toothbrush. There is no risk from kissing nor from sharing a Communion chalice; substances in saliva may inhibit the growth of HIV.

How long can AIDS survive outside the body? HIV has lipids in its membrane wall which protects it for several days against drying, but it does not withstand heat, and is easily killed by chemicals and by soap and water. If there is blood spillage, soak the area for thirty minutes in freshly made up Milton solution using a Milton tablet. Use disposable paper towels, and wear disposable gloves (preferably latex rather than plastic.) Seal all soiled items in a plastic bag and dispose of them carefully, preferably by incineration.

Is there any risk from being a blood donor? None whatsoever. Possible high risk groups, for example homosexuals and intravenous drug addicts, are asked not to become blood donors. All equipment is sterile and most is disposable, used once only. Sadly this myth discouraged many would-be donors.

Is there any risk of catching AIDS from a blood transfusion? In the United Kingdom, virtually none: all blood is tested for HIV antibodies. Because the antibody test does not become positive for a few weeks/months after infection, there is a remote risk, estimated at one in three million. Overseas, in countries where people sell their blood and where testing is patchy or non-existent, there is a significant risk.

Can AIDS be treated? Each year about 5 per cent of people infected with HIV will go on to develop AIDS; after five years 25 per cent and so on, with some claiming ultimately 100 per cent; leading a healthy life style may prevent or delay this happening. There is no cure for AIDS but each opportunistic infection which takes advantage of the body's depressed immunity is treatable. The costly drug Zidovudine can slow the progress of AIDS.

There are lesser forms of AIDS, namely, persistent generalised enlargement of lymph glands (PGL), for which there are many other causes, and which may get better, and also AIDS-related condition (ARC), which usually goes on to 'full-blown' AIDS and which causes symptoms like weight-loss, night sweats and diarrhoea, for which there are also many other causes, including anxiety. AIDS is often fatal in twelve to eighteen months.

Can AIDS be prevented? There is no vaccine, and once HIV is in the body, that person remains infectious for life. Health education is the means of stopping the spread of AIDS. As I hope this section makes clear, the vast majority of workers have no need to be personally concerned about AIDS. But it is important for everyone to be fully aware of the subject and to talk with other people who may not be as well informed, thus fostering this health education approach. In this respect AIDS is everybody's business.

Most people now realise that AIDS is not passed on in the workplace; they have more compassion for, and are more willing to work with, someone who has AIDS or is HIV-positive. At the same time their own life styles and heterosexual practices may not have changed – AIDS is still seen as happening to somebody else.

Shingles

Shingles – herpes zoster – is due to the same virus which many years earlier caused chicken pox and which has remained dormant in the body ever since. A blistering rash in the distribution of a nerve root is accompanied, and sometimes preceded by, pain which may be severe and/or persistent. Shingles can cause chickenpox in others and occasionally shingles in another adult, but contact with a child with chickenpox does not cause shingles in an older person.

Glandular fever

To most people 'glands' means lymph glands, but they may be salivary or hormone glands. Enlargement of lymph glands with or without a raised temperature is sometimes thought of as glandular fever, but there are very many causes of this. True glandular fever is a specific illness suffered once only, usually by young people, and it is caused by a virus which is confirmed by specific blood tests.

Glandular fever is not highly infectious, needs close bodily contact and has often been called the 'kissing disease', so there should be no risk of it being innocently passed on at work! Recovery normally takes a few weeks, but there can be longer-lasting debility.

Post-viral Fatigue Syndrome or Myalgic Encephalomyelitis (ME)

Viruses cause many different illnesses – the commonest is coryza, the common cold – which usually run an acute course, followed by complete recovery in days. A few virus conditions, however, can be followed by general debility which lingers on; in particular, this can happen after glandular fever or influenza. In ME there is general malaise and gross fatigue which usually gets better in weeks or months but can drag on, waxing and waning, almost indefinitely for years. In the end the condition clears completely. Although known as the 'post-viral fatigue syndrome', there is evidence of a persistent viral state. Fit young adults are most often struck down by ME. Other symptoms (all made worse by exercise) include muscle pain, headaches, lack of concentration, poor memory and tearfulness. As some or all these symptoms tend to affect everyone from time to time, there has been much debate in medical circles as to whether this condition is real. There are tests which prove the diagnosis, but a negative test does not rule out ME. There is no specific treatment and the best advice is to get plenty of rest, to eat well, and not to take on more physical or mental activity than can be coped with comfortably.

Infestations

The mere hint a member of staff has head lice or scabies causes consternation if the word gets out – everyone starts to itch! Either condition is looked on as highly contagious and a reflection of poor hygiene. In fact, both require close contact for transmission (for scabies, holding hands rather than shaking hands) and neither is necessarily related to dirt, failure to bathe or to change clothes regularly – lice *prefer* clean hair! An employee with head lice has most likely caught them from children at home who in turn picked them up at school. Communal use of brushes/combs in the staff toilet is a rare means of spread and the same applies to communal roller towels. Nits (the eggs) are stuck fast to the base of hairs and usually only elderly, infirm or injured lice are dislodged by brush or comb or washing and are no threat. A specific lotion treatment is rapidly effective.

Where it is the office which appears to be infested – be it with fleas, carpet, paper or book mites, flies or other insects from the foliage of plant displays, or fruit flies from a nearby catering unit, a specialist firm should be called in to apply specific disinfectant. Most flea infestations in this country are caused by the cat flea; the human flea is virtually extinct. Occasionally the bird flea spreads from a nest in the roof.

The flea population is highest in late summer. The flea can jump as

high as 23 cm. It feeds on blood and once replete hops off again.
Fleas are dozy in cold weather, but are kept fit by central heating.
Most bites are around the ankles, get through socks or tights, and
from a few hours to over a week later, cause anything from a tiny
itchy spot to a big swelling. It is the office rather than the person
which is infested, and expert help is required to track down and
eliminate the source.

Ringworm

This is a fungal skin infection, nothing to do with worms. Affected
areas may be ring-shaped and staff can pick up the fungus between
their toes in changing rooms and catering-unit-showers and then
transfer it via a towel to the groins. The condition is effectively
treated with specific anti-fungal creams.

DISORDERS OF THE NERVOUS SYSTEM

Epilepsy

About 2000 new cases of epilepsy are diagnosed in Britain each year
and there are over 250 000 people who are subject to fits. A GP with
the national average of 2500 patients will thus look after twelve
epileptics.

There are several varieties of epilepsy; 70 per cent start in childhood or adolescence; in most, no underlying cause is found and the
term 'idiopathic' is used.

Later onset suggests a specific cause, to be excluded in every case;
examples are a scar in the brain after a head injury, a brain tumour,
or bleeding in the brain. Over half of all fits occur at night and many
epileptics only get fits then. The next most frequent time is soon after
waking. In 'Grand Mal' epilepsy, the person may have a brief
premonition of what is to come, he or she then loses consciousness,
becomes rigid, teeth clenched, limbs stiff, then convulses, with
muscle groups alternately contracting and relaxing. This may begin in
one area and spread to the rest of the body, or be generalised from
the start. Breathing is affected and the face becomes congested.
Bladder and/or bowel control may be lost with involuntary evacuation. The fit lasts a few minutes, the convulsions gradually cease,

normal breathing is restored, the person wakes up feeling drowsy with a headache and often needs a short sleep. Sometimes unconsciousness is prolonged and occasionally one fit follows another; this is a medical emergency, justifying a 999 call, but usually one fit at work in a known epileptic does not warrant referral to a hospital accident and emergency department, rather that the person should be seen by their GP.

A 'Grand Mal' fit is alarming; the principle of first aid is to prevent self-injury, but not restrain. A pad, such as a padded spoon handle (rather than a pencil which will break), placed between the teeth before the fit starts will prevent tongue-biting; never force this as to do so risks breaking teeth. Loosen tight clothing.

After a fit at work, whilst the person may appear well again and want to go back to work, it is better for them to rest quietly in the first aid room and then be accompanied home, ideally to a relative (so as not to be alone at home). The GP should be contacted. For a time after a fit, though outwardly fine, the person may not be fully 'themselves' and have 'automatic' behaviour over which they have no control, and carry out acts which later they cannot remember.

Sometimes a fit occurs because the person stops the preventive drug treatment, either because they are feeling well and have been free from fits for so many months that the person considers further treatment unnecessary, or because the drug causes side effects such as drowsiness, or out of plain forgetfulness.

Complete control with suppression of all fits can usually be achieved using one or more drugs which should not nowadays cause serious side effects.

Fits are more likely when there is also a psychological illness like depression or anxiety. Psychological factors such as worry, excitement, grief and frustration may precipitate a fit, but this risk should not debar the epileptic from promotion. There is no need for the vast majority to live sheltered lives.

Once epilepsy is controlled, general management is by the GP who issues the prescriptions, but an epileptic should be re-assessed by a consultant every few years. Tests will show whether the optimum level of drugs in the blood is being achieved. The British Epilepsy Association provide an advice and information service.

The outlook for a person with idiopathic epilepsy is reasonably good. In 50 per cent of cases, attacks cease completely in time, and most of the other 50 per cent become fit-free with treatment. Full-time education and employment should continue, the person accepting the condition, understanding some restriction may be necessary, being prudent and not getting unduly upset about this.

Epilepsy in women may be aggravated by the premenstrual syndrome.

'Petit Mal' epilepsy is, in contrast, undramatic, is often not recognised and needs different treatment. The treatment for 'Grand Mal' makes 'Petit Mal' worse. Each attack of momentary blankness only lasts a few seconds. A school child with untreated 'Petit Mal' loses the thread of a lesson and goes to the bottom of the class. The adult at work is inefficient, errors are made and customers and colleagues cannot understand the lapses.

Other varieties of epilepsy in minor form may be incorrectly called 'Petit Mal'. 'Psychomotor' or temporal lobe epilepsy may be associated with behavioural change such as aggression.

Many epileptics regard epilepsy as a stigma and some may try to conceal the fact they have the condition, and be successful, unless and until a fit occurs at work. Some epileptics have experience of employers reluctant to employ them for fear fits will occur at work and endanger or upset the epileptic, customers and/or other staff. These employers also have the idea that attendance will be irregular with repeated sick leave needed. They do not appreciate that unless a fit occurs, the epileptic enjoys normal health.

So an epileptic, fearing rejection because of epilepsy, especially if this has already happened somewhere else, may not disclose the fact when next applying for a job and filling in the pre-employment health information sheet. A medical examination will not detect epilepsy and unless the history is given, there will be no reason to contact the GP. The person is then taken on, with the company unaware, until a fit occurs at work. The problem then, as with all previously undisclosed medical information, is what action to take. If there is a real risk to the safety of the epileptic and/or others should a fit occur, the answer is clear, otherwise it is not. Whilst the safety of the epileptic must be considered, this does not mean restriction to purely sedentary work. A job which involves working at heights is anadvisable, but not necessarily so a job requires occasional use of a step-ladder. An epileptic is eligible for the disablement register. Few jobs are barred, but they will include work where the lives of others would be at risk if a fit occurred. An example is a professional driver. Anyone who, for the first time, has an epileptic fit, should stop driving and notify the licensing authorities. After no fits for two consecutive years, with or without treatment, or if, in this time, fits only occur at night, the person may apply for restoration of a private driving license. A single fit after the age of seven makes that person permanently medically unfit to drive a heavy goods vehicle.

Headache

Headaches are the cause of much short-term sickness absence. Each year most people have a few headaches and 25 per cent of the population seem particularly susceptible. Enormous quantities of pain-relieving drugs are bought 'over the counter' and many of these are taken for headaches. There are very many causes of headache ranging from very serious to not so serious conditions. Examples are: *in the brain*: bleeding, blood clot, meningitis, growth; *relating to the head*: injury, sinusitis, tooth disorder; *to the eye*: glaucoma, conjunctivitis, or wearing the wrong glasses; middle ear infection; drug side effects, alcohol, fever, migraine, neuralgia, referred from the neck, tension/stress. The last cause is the commonest, but only an expert, that is, a doctor, can exclude the rest.

A doctor should be consulted if a headache persists or recurs frequently. It is one thing for someone with a headache to decide to take a couple of aspirin; quite another if those aspirin are prescribed by someone else, when the prescriber accepts the responsibility, especially if this is a lay person or a first-aider, who is not qualified to prescribe, not having been taught how to differentiate between all the various causes, for some of which aspirin or another pain relieving drug is inappropriate. This is one reason for not having any medication in a first aid kit at work.

Migraine

Migraine is due to blood vessel dilatation, preceded by spasm; typically it causes recurrent headaches on one side of the head and face, usually accompanied by temporary eyesight changes – seeing flashing lights, or partial loss of sight – and by vomiting. Migraine may run in the family. It affects 10 per cent of the population and comes after, rather than during, stress, regularly spoiling what had been planned for, say, the weekend. There may be premonitory mood disturbance, tingling of the fingers or altered vision. Migraine starts in the young: over 50 per cent of sufferers have their first attack before they are twenty, and 80 per cent by the age of thirty. Many more women than men are affected, in the ratio of three to one.

Migraine is helped by specific treatment taken early, combined with rest in a quiet, darkened room. Untreated migraine can last several days. Migraines can happen as often as three times a week. Headaches more frequent than this may be due to tension. Working on a VDU can trigger a migraine if the screen image is not stable, is too bright or has bright reflections.

A tension headache is described as feeling like a tight band or a heavy weight on the head: it is accompanied by anxiety, a worried look, a furrowed brow. Many people put down 'migraine' on self-certification when they mean 'tension'. Some people can stop migraines by avoiding certain food and drink. Migraines often fade away in middle age. If migraine is not disclosed in confidence on a pre-employment medical information sheet, mention in references of high incidence of sick leave in previous employment or at school will alert recruitment staff, and that information should be passed to the company doctor or nurse.

Multiple Sclerosis (MS)

Multiple Sclerosis (MS) was once also called Disseminated Sclerosis (DS). The pattern of this condition is very variable. Typically a person aged between twenty and forty develops an alarming nervous symptom like blurred or double vision, or tingling or clumsiness in one limb, which soon clears and may never recur but more usually does, with a pattern of relapses and remissions and gradual deterioration into disability. A remission may go on for years but each tends to be less complete than the last. There is the view that becoming overtired or highly stressed hastens a relapse, which has implications for the sort of work someone with MS should undertake. Treatment can shorten a relapse but not prevent a recurrence. Sick leave during a relapse is usually necessary, but in remission, even if there is disability, the person with MS should if possible keep working – sedentary work is suitable if within easy reach of home accessible by, and with facilities for, a wheelchair when and if this becomes necessary.

Parkinson's Disease

From middle age on and more commonly in men, there is insidious development of stiffness and tremor spreading from one hand, with

Cancer starts with one rogue cell 'going mad' and repeatedly growing and dividing; why this should happen is often still unknown. Almost any tissue of the body can be affected, some much more commonly than others.

Prognosis depends on the characteristics of the cancer cell, the extent of local and distant spread at the time of diagnosis, and the response to treatment. Often this treatment is surgical, to remove the primary growth totally. Spread, via lymph channels and the blood stream, causes secondary growths. Radiotherapy and cytotoxic-cell-poisoning drugs destroy secondary growths and are now being given early, when there is a better prospect of eradication of the tumour. Logically, the earlier treatment is given, the better the chances of success, although apparently identical cancers behave very differently in individual people. Hence the move towards screening the healthy population for presymptomatic detection, or to find the transitional phase before a cancer starts, as in smear test screening of women for such cell change in the cervix (see Chapter 6).

The ideal approach is of course the preventive one, which involves the elimination of the factors which cause cancer.

On sickness certificates, cancer is disguised as Neoplasm (NG) (New Growth), Ulcer or Tumour. Nowadays, patients are more likely to be told the whole truth if they want to hear it, but, for example, cancer of the bowel diagnosed at operation may at first be referred to as 'an ulcer in the bowel wall' which has been removed. More publicity is needed to show that cancers can be cured, particularly when found early and dealt with promptly. Cancers on or near the surface of the body can be identified at once, hence the value of regular screening for breast cancer as well as for the changes in the cervix that lead to cervical cancer.

The outlook is poorer for cancers which become apparent late, in the stomach or lung, for example.

With regard to employment, whilst it is difficult to generalise, extended sick leave is usually necessary for the initial surgery and for full recovery from this and any radiotherapy or drug treatment which, as stated, is usually given early rather than only if 'secondary' growth occurs. Some cancers are speedily dealt with, an example being an early, localised, skin cancer. It depends on the degree of side effects, whether sedentary work is possible during the repeated courses of drugs which are often given in an attempt to prevent secondaries. Debilitating side effects are usual during and after radiotherapy, but usually this is a single course. Regular follow-up examinations are necessary to ensure there is no recurrence, either near the original growth or distant from it. The interval between examinations is gradually extended; there is said to be clinical cure after five years free from the disease. Keen as patients may be to get back to work, it is usually better for them to wait until the course of treatments is completed.

In cases where the cancer has spread, sick leave should if possible be extended, and retirement not mentioned by the employer until the employee realises return to work will be impossible; even then, if death is not far off – for instance, within twelve to eighteen months – 'death in service' is usually the kindest and best option.

Kidney failure

The two kidneys have a fantastic reserve, but if kidney function falls to less than 5 per cent of normal, then regular dialysis or/a kidney transplant is necessary to clear the body's soluble waste products. Supply of kidneys nowhere near meets demand, so there are criteria for selection; one is an age barrier of being under forty-five years old. Dialysis takes ten hours, three times a week, first in hospital and then where possible at home. Two-thirds of patients are suitable for home dialysis, when night-time dialysis will allow a normal working week. 90 per cent of home dialysis patients get back to full-time work. Six to twelve weeks' training at a hospital dialysis unit is needed, and in this time home dialysis equipment is installed. A relative, who also requires training, must be available to help set up and run each dialysis session, using machinery which is expensive to buy, instal and maintain. A somewhat simpler alternative is peritoneal dialysis. Adapting to the restrictions imposed on life style takes some getting used to; also there is the strain of the uncertainty of the future affecting the patient and close relations. Five-year survival on dialysis is 80 per cent.

The alternative is donor kidney transplant and a successful transplant allows return to a virtually normal, unrestricted life.

Diabetes

Diabetes results from pancreas failure, with the result that there is little or no natural insulin to regulate blood sugar. Diabetes can happen at any age. In the young, the onset is sudden; in the older person, it is more gradual, over months, and diagnosis may be missed until a problem occurs like a skin infection which will not heal. The person may have none of the classic symptoms of thirst, weight loss and passing lots of urine. Unsuspected diabetes is also detected when sugar is found in the urine during a health screen.

A diabetic on treatment can live a normal life, but however well diabetes is controlled, complications may eventually occur affecting, for example, the kidneys and the eyes. This should not deter a company from employing a diabetic, although a few jobs would be unsuitable, including those with a changing shift-work pattern. Most young diabetics need insulin twice a day by injection, whereas in the older person, change of diet, losing weight and sometimes tablet treatment may be all that is necessary. Where insulin is needed, the diabetes is stabilised in hospital. All diabetics need regular visits to a diabetic clinic or GP hospital-shared scheme. Day-to-day self-monitoring is carried out by regular blood tests, using a small portable instrument which gives a reading of the glucose level from a drop of blood.

Occasionally diabetes is 'unstable'; blood sugar and thus insulin dose fluctuate widely; there is risk of a low blood-sugar state – hypoglycaemia, 'a hypo' – the opposite of the high blood sugar with which diabetes presents. Usually there are warning symptoms of pallor, restlessness and sweating, and at this stage eating a few lumps of sugar or drinking sweet tea will restore the blood sugar quickly before the onset of abnormal, irrational behaviour, poor concentration and unsteadiness. In the absence of treatment, unconsciousness follows. Some diabetics prefer not to mention their condition, but because of the possibility of hypoglycaemia, it is important that others around them are aware of the warning signs and know what helps. If warning signs are spotted early, the cup of sweet tea or a few lumps of sugar (which the diabetic should always carry, together with a card giving the diagnosis and treatment) is effective, but once unconscious treatment must be given by injection by a doctor. At the end of the initial hospital in-patient stabilisation process, the diabetic person was taken towards a low blood-sugar state to learn awareness of the warning symptoms, but as 'hypos' damage the brain they must be avoided.

Stoma

The bowel opening has been surgically transferred to the abdominal wall, where it opens into a bag. The prospect of having a stoma is distressing: the sight of this bright red, contracting bowel-end takes some getting used to, but for a number of people of all age groups, a stoma is a temporary or permanent necessity for a variety of bowel disorders.

A *colostomy* is a large bowel opening and an *ileostomy* a small bowel opening. Specially-trained nurses advise on stoma care. Voluntary help is available from the Ileostomy and Colostomy Associations.

When the stoma is temporary and part of curative treatment, acceptance is easier and more total than where the procedure is permanent or palliative. People with a temporary stoma are inclined to treat the time with it as an interlude. However, a few patients do not fully accept the stoma, being fearful that the specialist may change his or her mind and make the stoma permanent. The disposable bags and their fittings are easy to put on, undetectable when the person is dressed, and leakproof. There is no odour problem. Provided there is the ready availability of a lavatory where the stoma bag can be changed – a simple, quick procedure – there should be no difficulty at work.

Despite all this, many stoma patients (ostomists) expect that they will have to give up work altogether, or at least seek a part-time job. Sadly, despite improvements in the appliances, and encouragement to lead normal lives, surveys still show there is the strong likelihood of changes in life style and increased isolation. Ostomists feel they will not be accepted as normal, healthy individuals by their family, work colleagues and friends.

Irritable bowel syndrome

Irritable bowel syndrome (IBS) otherwise known as spastic colon is very common. It may also be referred to as spasm or cramp of the colon. It causes bouts of chronic abdominal pain and/or loose stools or constipation. IBS is the most common diagnosis made in cases referred to any hospital gastroenterology out-patients' clinic. It is important that anyone, an older person in particular, who has any of these symptoms which persist, should consult their GP, as not only is treatment and advice available if the diagnosis is IBS, as is most likely, but also a more serious condition must be excluded, principally cancer of the bowel, for the earlier that is diagnosed, the more likely is cure. The other significant bowel symptom which does not occur in IBS, but which needs immediate attention, is blood in the stools. The most likely cause is piles, but once again, a more serious cause must be excluded. The term 'bowel' is synonymous with guts or intestines, which are divided into the small and large intestine. IBS affects the large intestine, otherwise called the colon. In IBS, there is tightness and tension of the colon which causes either diarrhoea or constipation. If the overactivity is generalised, the contents of the colon are rushed through, whereas more local spasm will cause hold up of the contents and constipation.

The colon works best if it is reasonably full, so it prefers bulky foods containing plenty of fibre to work on. Thus the mainstay of treatment for IBS is high-fibre foods. If symptoms persist despite this change of diet, the general practitioner may prescribe anti-spasm drugs. Occasionally, a person with IBS is suffering from depression and needs specific treatment for this which helps the IBS also. If nothing works and organic disease has been excluded, psychotherapy or hypnotherapy may help.

Medical investigations such as X-rays show structure rather than function, so in IBS the result of investigations will be normal, the problem being abnormal function of the large bowel. Thus a negative outcome of investigations does not mean there is nothing wrong – symptoms are very real, the pain, the distension and altered bowel habit being linked to spasm and distention.

Though chronic, IBS is benign and does not progress to anything sinister, like cancer or colitis. Any foods that seem to aggravate it should of course be avoided, but I do not think IBS is based on food allergy. Stress can aggravate the symptoms, and the practice of relaxation techniques can be effective. Calm people have calm digestive systems. The symptoms of IBS feature on sickness-absence record cards as the repeated cause of short-term absence, and there will also be absences for investigations.

16 Aspects of Women's Health

In this country, women's patterns of work have changed in the past twenty-five years. More women work and there are better facilities for continuing a career whilst raising a family. Five million of the nine and half million women in employment are aged between sixteen and forty-four. Over a quarter of mothers of children under five are in part-time or full-time work. More women with children of any age are returning to work.

A survey of part-time work by Robertson and Briggs of the Unit for Manpower Studies shows that about 40 per cent of the work force are women, and about 40 per cent of these women work part-time, 78 per cent of them are married and they are mostly in the age range of twenty-five to fifty-four.

A woman's health can be affected by physiological changes related to menstruation and pregnancy, although this is by no means inevitable. There are also gynaecological and breast conditions she can get which will worry her.

PREMENSTRUAL SYNDROME

The term 'premenstrual tension' suggests just one thing, tension, so the condition is better called Premenstrual Syndrome (PMS) to encompass the whole range of symptoms which include, besides tension and irritability, undue tiredness, depression, breast discomfort, headache, sickness, backache, and fluid retention (the abdomen feeling bloated, and swollen ankles). One or more of these symptoms can start as early as fourteen days before the period is due, when hormone levels change. Symptoms clear completely soon after the start of the period, only to return the next month. PMS affects up to 50 per cent of women who, whilst affected by PMS, are likely to be less dexterous, more accident-prone and more emotionally volatile.

PMS can create difficulties at work and at home and 'rubs off' on the husband, whose work may suffer in consequence; marriages break up because of PMS.

Work efficiency can suffer and short-term sickness absences occur at this time. The possibility of PMS can be judged from scrutiny of the sickness-absence record card, that will show a monthly pattern. PMS either develops when menstruation first starts or after childbirth – freedom from symptoms during pregnancy suggests the diagnosis.

Hormone imbalance, in particular a lowered level of progesterone, appears to be responsible for some cases of PMS. Hormone therapy or vitamin B6 (Pyridoxine), during the premenstrual phase is effective in many of these cases. In at least half the women affected, the diagnosis is never made, usually because they do not consult their GPs. For them the monthly pattern that can upset all around them, goes on until the menopause.

DYSMENORRHOEA

Dysmenorrhoea (the tongue-twisting title, usually shortened to 'dysmen') is the medical term for painful menstruation or 'periods', the monthly blood loss some women refer to as the 'curse'. From the menarche at around twelve, menstruation continues to the menopause, with gaps during pregnancy and breast-feeding. Menstruation is also affected by a number of physical and psychological conditions.

Dysmen is one of the commonest gynaecological complaints in the civilised world, and is most prevalent in young, single women who lead predominantly sedentary lives; office work comes into this category. When severe, it is a reason for short-term sickness absence, causing incapacity for the first one or two days of each period.

The true incidence may be disguised because of reluctance to disclose the true cause to a male manager. Instead, a miscellany of other reasons are given such as 'tummy upset' or 'sickness'.

Of the three varieties of dysmenorrhoea, congestive, spasmodic and membranous, spasmodic is the commonest; at least half the female population suffer this to a varying degree at some stage in their lives, although under 10 per cent seek help from a doctor. There may be considerable discomfort the day before the period starts, then, at the start of the period, intermittent cramp-like pains in the pelvis and lower tummy (which can be excruciating for up to an hour at a time), radiates to the back or thighs and is accompanied by nausea, sickness, diarrhoea and/or faintness. This is followed by less severe pain for up to twelve hours. Symptoms are worst during the heaviest bleeding.

Spasmodic dysmenorrhoea first starts in the early or late teens, is usually most severe between the ages of nineteen and twenty-one, and is rare over the age of thirty-five. Marriage often, and pregnancy usually, ends the condition.

In contrast, congestive dysmenorrhoea starts later in life, is present before the period and is always relieved by the start of the period. Associated symptoms are unusual. There is often an underlying gynaecological disorder, though the condition may be related to sedentary life.

It has been claimed that two personality types are especially susceptible to dysmenorrhoea: the immature, shy person using menstruation as a way to escape difficult situations, and the aggressive tomboy, resenting periods.

Simple preventive measures help. They include:

(1) Regular exercise.
(2) Regular bowel movements.
(3) Not thinking about the next period.

There is a range of effective treatment available from doctors if simple pain-relief tablets and a hot bath are not enough. There is no need to just put up with this condition.

PREGNANCY

The British birth rate has for many years been in decline, for such reasons as modern contraception, legalised abortion and the cost of living.

Chemical confirmation of pregnancy, if needed, is quick and reliable. The date of the baby's birth, the expected date of delivery (EDD), is approximately nine months and one week after the first day of last menstrual period (LMP), e.g. LMP 3 February, EDD 10 November.

Pregnant staff are entitled to benefits from both State and (provided certain conditions are fulfilled) their employer. Under the National Health Service, the state provides antenatal and postnatal care.

Most births are in hospital; with a normal birth, length of stay ranges from forty-eight hours to ten days.

Antenatal examinations start at three months into pregnancy and continue monthly to thirty-two weeks; once a fortnight to thirty-six weeks, and weekly thereafter, to 'term' at forty weeks. Provided there are no complications, these may be carried out by the GP or alternate between the GP and the hospital clinic. The mother-to-be may want to skip these checks, especially if she feels well, if there is a long wait at the clinic, and a lot to do elsewhere, but it is important for the future well-being of mother and baby that all appointments are kept.

The state gives a Maternity Grant for which most women are eligible.

Pregnancy and employment

There is no proof of any risk from normal office work during pregnancy to the health of the mother or baby; a few women work right up to the expected date of delivery (EDD).

An employer must allow time off for antenatal appointments with a doctor, midwife or health visitor throughout the pregnancy.

From 6th April 1987, a Section of the 1986 Social Security Act covering Statutory Maternity Pay (SMP) came into force. As from June 1987 SMP has been paid for up to eighteen weeks, starting between the eleventh and sixth week before the EDD, to any pregnant woman who has been continuously employed for at least six months by the fifteenth week before the EDD and who stops work for this time – the maternity pay period (MPP).

In some countries there are more extensive maternity rights: for example in Sweden there is provision for 270 days paid leave, 90 of which can be taken by the father.

A woman needs to apply in writing for maternity absence at least twenty-one days before the date she intends to stop work, and also at the same time needs to indicate her intention to return to work – if this is her wish. (Employment Protection Provisions relating to the right to return to work after childbirth require a previous period of two years continuous service of more than sixteen hours per week or five years continuous service of between eight and sixteen hours per week, in order to qualify.) She will need to provide a valid maternity certificate (MAT B1) issued by a doctor or midwife not earlier than fourteen weeks before EDD. Return to work can be at any time before the end of twenty-nine weeks from the week in which the birth occurred, provided the woman gives at least twenty-one days' notice in writing, which can later be postponed for up to four weeks either by the employer or by the woman because of illness, supported by a doctor's certificate.

Promoting good health in pregnancy is assisted by providing the following type of health information, issued by the company doctor or nurse. The example below was in part written by a former National Westminster Bank Medical Officer, Dr Marian Roden:

MEDICAL ASPECTS OF PREGNANCY

Although pregnancy is not an illness – indeed it is a time when you should experience an increased sense of well-being – it is traditionally managed by health care professionals, principally Doctors, Midwives and Health Visitors. From the moment of conception onwards your baby is developing inside you and is dependent on your body for all that he or she needs to grow from a tiny fertilised egg cell into a full-term foetus with complex organ systems, ready to survive outside the protective environment of the womb. It is therefore of paramount importance that you look after yourself and keep your body in tip-top condition throughout the course of your pregnancy.

The following guidelines should help to ensure a happy and healthy pregnancy with a bonny baby at the end of it.

Antenatal Care

Report to your own Doctor as soon as you think you may be pregnant. It is important that antenatal care begins as early in pregnancy as possible so that you obtain good advice from the outset. Your Doctor will arrange for you to have regular check-ups throughout your pregnancy so that if any problems do develop they can be diagnosed and treated at an early stage. Furthermore, this will mean that you will have plenty of opportunity to discuss any queries or worries that you may have had. You will also be given information about classes for parents-to-be run in your neighbourhood. The broad spectrum of antenatal care thus ensures that your pregnancy is watched over by professionals and that *you both* learn all you need to know about pregnancy, childbirth and parenthood. It is geared to give your baby the best possible start in life and to prepare *you both* for your role as parents.

Leaflets on various aspects of pregnancy are available free of charge from the Health Education Unit of your local Health

Authority or from the supplies department of the *Health Education Authority*. Your attention is drawn in particular to 'The Pregnancy Book' which offers a complete guide to becoming pregnant, being pregnant, your rights and benefits and caring for your newborn baby. If within your present working unit or nearby the Bank employs a Nursing Officer, you should not hesitate to make contact for help or guidance regarding your pregnancy, eg information, explanation of medical terms and guidance on nutritional matters.

Diet
Provided that normally you eat a mixed healthy diet, there is no need to make any drastic changes during pregnancy – in particular you don't need to eat for two! During the course of your entire pregnancy you should gain about 25lb in weight, most of which is put on during the last twenty weeks. If you are seriously over or under-weight when you become pregnant, your Doctor will advise you how to correct this, since either condition can give rise to problems for both you and your baby. Furthermore, excessive weight gained during pregnancy is difficult to lose afterwards and may lead to permanent obesity.

Further tips
Smoking – don't! Smoking is bad for you anyway and it is now established beyond doubt that mothers who smoke during pregnancy give their babies a poor start in life. So give it up, preferably before you conceive!
Alcohol – a little if you must, but preferably none at all.
Teeth – your teeth are particularly vulnerable during pregnancy so be sure you receive regular dental care – it's free during pregnancy and until your baby is twelve months old.
Exercise and rest – plenty of both.
Pills and Medicines – none except those approved or prescribed by your Doctor.

In normal pregnancy the mother-to-be is usually well. If there has been a previous history of miscarriage, she may be advised to stop work and rest from early pregnancy on.

The most frequent symptom in the first three months of pregnancy is 'morning sickness' – early morning nausea with or without actual vomiting. If this is no more than a temporary inconvenience, her doctor may not prescribe, and a woman anyway may not want to take, medicine to relieve this symptom. This is particularly so since the thalidomide tragedy. If the sickness is more troublesome, there are a number of effective and safe drugs. Fortunately, the condition is usually not severe and almost always clears within, or soon after, the first three months. Occasionally, if very severe, admission to hospital is necessary.

The employer should not expect a pregnant employee to stand for long periods or to lift and carry heavy weights.

When a woman ceases to work during pregnancy will depend on her health, the health of the unborn child, the nature of the work, her own wishes and those of her partner. Similar conditions (such as her recovery from childbirth, the health of the baby, and so on) govern how soon she returns to work after the birth.

For many months after the birth, though the mother is fully recovered, her baby is, of course, still very dependent on her, particularly when ill or if being breast-fed.

BABY CARE

The difficulties of working and at the same time looking after young children could be stressful. People in this situation may be the sole breadwinner of one-parent families or a mother who wants to continue a career. As many as one in ten children aged under five are looked after during the working day by child-minders.

There are about 32 000 registered child-minders in Britain and many more unregistered. The National Child-minding Association has produced leaflets on finding and being a child-minder.

A registered child-minder may look after a child in her home for up to eight hours a day.

Private and state nursery schools take three to five year-olds during normal school hours. There are also local authority day nurseries and pre-school playgroups. There is often no provision for care during school holidays. A number of successful self-help local community care centres have been set up, some with the help of local authority grants.

A few employers provide a crèche at the work place staffed by trained children's nurses. There can be enormous practical difficulties – bringing small children on busy commuter trains for example. An alternative is for two mothers, each with a young baby, to share looking after, and sometimes breast-feeding, both. Another alternative is job-sharing, with the two people who share the job spending the time not at work looking after both sets of children. Whilst slow to catch on in this country, jobs have been successfuly shared in the

United States and Scandinavia. The Equal Opportunities Commission and the Manpower Services Commission have recommended research into the potential of job-sharing. Level of responsibility and promotion prospects can be greater than for part-time work. The employer has the experience and expertise of two people for the equivalent of one person's salary; also, confident their children will be well cared for, both members of the job team can concentrate fully on their work.

GERMAN MEASLES IN PREGNANCY

German Measles, rubella, is a mild virus infection with an incubation period averaging eighteen days; it is infectious for the few days before and after the onset of the rash and is spread by droplets from coughs or sneezes, or by direct contact; the rash is pink, and flat on the skin; there are associated enlarged neck glands and sometimes, temporarily, pains in the hands and feet.

A not-infrequent cause for concern is when a female member of staff has been working in the same office as someone who has rubella, thinks she has never had German Measles herself, and either is pregnant or has been in contact with someone who is pregnant.

The concern is that in the first four months of pregnancy, should the pregnant woman develop rubella, there is considerable risk of the unborn baby also contracting the illness, and whereas in the vast majority of children and adults, rubella is a fleeting illness of no severity, the unborn baby can be seriously damaged. Between 200 and 300 rubella-damaged babies are miscarried or stillborn each year. The frequency of abnormalities, particularly affecting the eyes, ears and heart, decreases from conception to sixteen weeks of pregnancy (50 per cent in the first month, 30 per cent in the second, 15 per cent in the third, and 4 per cent at sixteen weeks).

Many rashes are incorrectly called rubella which accounts for some people's claim of having had two attacks – life-long immunity should follow a single attack.

Fortunately, rubella can be prevented by one injection of live, attenuated rubella virus vaccine (the virus is altered to stimulate immunity without causing the illness). Rubella vaccine is safe but it is clearly essential not to have vaccine during pregnancy and to avoid pregnancy for three months afterwards. Rubella vaccine is now offered to all school girls between the ages of eleven and thirteen; girls who miss that opportunity can be screened and vaccinated if necessary by their GP, as should unvaccinated women considering pregnancy who have not had rubella. The need for all this and the incidence of rubella will fall following the introduction of a triple vaccine MMR – measles mumps and rubella – which is given in infancy and to boys as well as girls.

Unvaccinated pregnant women exposed to the virus should have a blood test for antibodies to show whether they have unknowingly had rubella in the past. If they have not, a further blood test after a three-week interval will prove whether, in that time, they have caught rubella, in which case the antibody level will have gone up. If it has, termination of pregnancy is recommended to prevent a possible catastrophe.

MISCARRIAGE

Both abortion and miscarriage mean termination of pregnancy, either spontaneously or induced before the twenty-eighth week of pregnancy. 15 per cent or more of all pregnancies end in spontaneous miscarriage; in 60 per cent of these, something is wrong with the developing baby. Commonest during the first few week of pregnancy, spontaneous abortion is sometimes unsuspected and regarded as a late period. When there is abnormality of the developing baby, a 'threatened' abortion always becomes 'inevitable' within the first sixteen weeks of pregnancy, and a small gynaecological operation (D & C) may be required.

Treatment for threatened miscarriage is principally by rest at home or in hospital.

INFERTILITY

A tenth of all marriages are childless, many deliberately so. Of those unable to conceive, investigations in a third show the cause is the husband, and the wife is fertile.

MENOPAUSE

The menopause signals waning production of the female sex hormones. The most obvious sign is that the monthly periods stop. This may happen suddenly or gradually.

To refer to any other symptom as menopausal is strictly incorrect, as menopause simply means the end of menstruation. The transitional four to five years around the menopause, during which reproductive function declines and symptoms occur, is called the climacteric.

The menopause occurs on average in the United Kingdom at age fifty-one. The accompanying climacteric generally occurs from age forty-nine to fifty-three. It may be a smooth, scarcely noticed transition, or cause temporary symptoms, or be a prolonged difficult time, physically or psychologically or both.

Embarrassing, devastating hot flushes and night sweats are physical symptoms which affect about 75 per cent of women and continue in 25 per cent of cases for more than 5 years. Depression, apathy, irritability, anxiety, loss of self-confidence and poor concentration are common psychological symptoms. Hormone replacement therapy (HRT) should abolish these symptoms but brings back periods. Menstrual-type blood loss, not linked to HRT, and occurring more than six months after the menopause, termed post-menopausal bleeding (PMB), though usually of no significance, must be investigated to exclude cancer of the womb. HRT also protects against thinning of the bones, osteoporosis.

HYSTERECTOMY

Each year in the United Kingdom 60 000 women have a hysterectomy. In some cases, besides removal of the womb, also removed are either or both ovaries (the sex-hormones-producing glands). Ovarian cancer is often diagnosed late and causes 3000 deaths a year, more than any other gynaecological disorder, but preserving the ovaries avoids the sudden, menopausal symptoms which occur if both are removed. Before this or any other surgical procedure, the patient should be fully briefed by the surgeon and with time for fears to be fully aired and myths dispelled – for example, femininity is unaffected. An average of ten days is spent in hospital. Though the woman may straight away feel better than she has for years, it can take up to six weeks before improvement is noticed and up to three months to get back to feeling one hundred per cent and able to resume full activities (though it should not be necessary to stay off work all this time if the work is sedentary).

BREAST CANCER

Often nowadays the treatment of breast cancer involves the removal of just the growth (lumpectomy) followed by X-ray and drug treatment, rather than the psychologically traumatic removal of the whole breast (mastectomy), with or without the surrounding lymph glands through which the cancer may spread.

Many women have great difficulty in adjusting to what has happened (around 25 per cent become depressed within a year of mastectomy), especially when before the operation the diagnosis of the lump was not known for certain.

However, surveys show that three women out of four come to terms with the loss of a breast within two years of the operation. A prosthesis helps – an artificial silicone breast, either external or inserted under the skin. The Mastectomy Association, founded and run by women who have had this operation, provides invaluable support.

Other treatments, used with or without surgery, to destroy any locally-recurring cancer cells, include radiotherapy (X-rays shone on to the breast area which also cause sunburn-like skin changes and temporary general tiredness and nausea), and chemotherapy courses of drugs which can damage healthy, dividing cells as well as rapidly killing cancer tissue (during the course of treatment there may be tiredness, sickness and temporary hair loss). One drug increasingly used long-term to prevent recurrence, Tamoxifen, is largely free of these debilitating side effects.

17 Aspects of Mental Health

STRESS

'Stress' is a hackneyed, imprecise term, confusingly used to denote both cause and effect, better described as strain. Stress is the subject of countless talks, articles and books and is blamed as the cause of much mental and physical ill health, so it may seem logical to conclude that stress is to be avoided at all costs. It is sometimes used as an excuse. Change, which is at the root of much stress, happens faster than fifty years ago, and life generally has a faster pace especially in communications and travel, but stress has been present in one form or another since life began. To blame the increased incidence of heart disease or nervous illness on stress alone is too simplistic, and incorrect. In essence harmful stress arises when resources and demands are out of balance; demands being either too little or too great. Another analogy is that the stressed person feels unsafe.

Stress gets a very bad press and it cannot be emphasised enough that some stress, better termed pressure or challenge, is very necessary to bring out the best. Life without stress would be very dull – a vegetable existence – and is anyway impossible: you cannot escape all stress; for a doctor to recommend a stress-free job is just 'not on'. Young staff should never be hesitant about taking on added responsibilities within their capabilities, for fear they may later have, say, a heart attack or breakdown. If a graph is drawn of stress load versus performance, initially an increase in stress correlates with improved performance until a plateau is reached. If the stress load continues to increase, and causes distress, performance then starts to fall away, followed by a breakdown in health with physical or mental illness.

Stress from any source, domestic as well as work, is relevant. It is the accumulation of personally-significant stresses which matters. In the face of too much happening and tending to go wrong, everyone has a breaking point; but illness, whether psychosomatic (such as varieties of eczema and asthma), or mental, is more likely in the vulnerable 'neurotic' personality, particularly the perfectionist obsessional, who is stressed by events which normal people take in their stride. In the months before the breakdown there have often been a series of personally-significant, stressful events at work or at home, or most likely, in both places at the same time. In the United States, scaled lists of life-events stress ratings have been produced, with each event given a score between one and a hundred, based on questioning many people about which life events matter most to them. Death of spouse or divorce score the highest at one hundred, and minor violation of the law, together with Christmas, the lowest at twelve each. Christmas is included possibly because people are then stuck with relations they detest. A score over one hundred and fifty for the previous six months puts that person at increased risk of illness in the following six months.

Most stresses involve change or loss. Not all are bad. Good news features as well. An example of a risk situation is promotion or a change of job which involves moving house. At work there are new colleagues to get to know and the new job to learn and shine in; this is more stressful if the previous job-holder was a success and set a high standard to live up to. On the domestic front, there is the problem of house-hunting, buying one home and selling the other, then the move itself. The employee may be staying in a hotel from Monday to Friday, separated from spouse and family, bored and lonely with nothing much to do in the evenings but eat, drink and watch TV – an unhealthy life style, especially when combined with a full English breakfast each day. (Over the years this great English breakfast, high in fat, may contribute to the great English heart attack.)

Stresses on the family will affect the employee, too. The spouse has been uprooted from friends and neighbours and may have had to give up a job she or he enjoyed. The children also lose friends and the new school may have a different curriculum – of crucial importance in examination years. Stress is also likely if the new job is overseas, especially if a language barrier is added to culture shock and change of climate for the whole family to adapt to. Courses are available to ease this transition. Not to be overlooked is the stress of readjusting back to normal life in the United Kingdom after two or three prestigious years in an overseas capital as 'Our man (or woman) in . . . .'.

A stress-related illness, particularly a nervous illness, is more likely to affect the person with deadlines to meet, who is unable to delegate or even worse, cannot say no, and, like a sponge, takes on the work of others as well. A stress-related illness, particularly when this is a nervous illness such as anxiety and/or depression, once it happens, lasts for weeks, or months. When the illness is at its height, it is impossible to predict accurately how long it will be until there is sufficient recovery for a return to work. When any psychological illness is bad enough for sick leave, prediction of how long that sick leave will last is particularly difficult. Work schedules are disrupted and, as with any illness, the manager has to make the decision whether to find a temporary replacement, appoint a permanent successor, or ask the other staff to cope and absorb the extra work in the hope of an early return.

With hindsight, others may have seen a mental illness coming for a long time. But how, a manager may ask, can I be sure to spot a person who is heading for this kind of illness, so I can look for any work-related factors and guide the person to their doctor in an attempt to prevent an illness developing? What I need is a checklist. Identification is not that straightforward. In some cases there is obvious abnormal behaviour, in others it is only possible for the skilled trained observer to detect abnormal behaviour patterns in the early stages. Early signs include irritability, loss of a sense of humour, and disturbed sleep. Concentration and decision-making falter. When the mental health of the member of staff causes concern, the questions to be answered are:

(1) What is happening?
(2) Why is it happening?
(3) Can I do anything to help?

Doctors deal with the first question, but a manager may be able to help by telling the doctor in what way behaviour has been abnormal. 'Why' involves consideration of personality – a person who has coped successfully with such stresses as school, puberty and marriage is more likely to have a physical illness causing the behaviour changes than a stress-related mental illness.

A doctor will get clues to a diagnosis of depression with a few simple questions on behavioural changes as contained in the copyrighted 'general health questionnaire', and covering such areas as sleep, mood and work pattern.

People at risk

Psychological illness causes some of the longest periods of sickness absence, and involves suffering for both the individual and their family. It may deprive the employer of the services of the employee for many weeks or months. Sometimes the illness is blamed on the job alone, but stresses outside work are often as relevant as those at work. However, staff may resent any enquiry – 'What business is it of yours?' is a typical rejoinder; but where a manager has developed rapport with staff, it is unlikely that sympathetic enquiry will be rebuffed. He or she is also likely to know which staff have chronic insoluble domestic problems. It is these people who are particularly at risk when there is, in addition, a personally-significant stressful life event. There are limits to what anyone at work can do when a colleague seems to be becoming mentally ill. If the manager or a colleague speaks to the personnel department or senior management, the individual concerned may look on this as being 'shopped'. He or she assumes that any mental illness will be looked on as a sign of weakness and may mean the end of future promotion prospects, with the permanent question mark 'Can he/she take it?' added to the file. If management is aware of what is happening, in some cases temporary alteration of the workload, and advice to see the GP, plus the willingness to listen, may avert a full mental illness developing. Access to skilled independent confidential counselling such as by the company nurse may make all the difference.

It is obvious that people work better if contented and happy with their jobs. There is a risk that morale will fall as a company grows, if this leads to anonymity with loss of that sense of 'belonging'. Factors contributing to job satisfaction include full use being made of individual abilities; interest in the work done and the ability to choose to some extent what work to do and what to turn down. It clearly will help to be adequately rewarded for work well done. Particularly relevant to this book is a finding that secretarial and clerical workers were less satisfied with their jobs than those in other occupations.

Prolonged misemployment, for example the classic 'square peg in a round hole,' with all the attendant frustrations and insecurities, has an adverse effect on well-being and long-term health. Occupationally speaking, thwarted ambition, dashed hopes, unwelcome responsibility or long-term job insecurity may also harm health.

Continuing this personnel nightmare, other likely at-risk groups are those who have no clear job description and no defined limits to their jobs; also the underemployed; the overpromoted; those who have missed out on promotion or the opportunity for promotion; and those who have an unsatisfactory working relationship with their boss. A type 'A' personality, highly time- and deadline-conscious, aggressive and angry, is more at risk than a more laid back type 'B', as he or she feels much more pressured by what he or she sees as overwork.

Even the most stable, healthy and committed staff will become demoralised when working for a company which appears to do little or nothing to encourage or reward their efforts and which harshly criticises any shortcoming. It is not just absence of praise which does harm, but absence of any significant feedback, an apparent indifference.

Some people are stress creators and have the knack, whether by design or 'it just happens', of upsetting others. A pattern can be detected and management action is needed to deal with this. Pointers are: the person interrupts others; is rude and abrupt; has no real interest in other people; will not let them get on with their work; is ready to criticise, and is aggressive and very hard-working him or herself.

The relative significance of the different work stresses depends to a certain extent on the employee's level in the company. For example, if there is any uncertainty about the company's future the senior executive will be more concerned than the middle manager, who is more concerned about his or her role within the same company. All too often, it seems, middle management is squeezed between over delegation by top management and 'shop-floor' militancy. Common to any unsatisfactory employment situation are poor communications, including lack of feedback, inconsistency, and insensitivity; there may also be disparity between the task and the time allocated to complete it; also between the quality of work personally desired, the quality expected by others and the quality achieved. This particularly hits anyone with a perfectionist/obsessional personality. Some common examples are: sudden change in hours of work or in working conditions and/or responsibilities, with little or no warning or explanation. For instance, a manager on the verge of a nervous breakdown returned to work after the old-fashioned, recuperative holiday 'away from it all' to find his job quite different. This was without any warning. There was no apparent concern or even enquiry from his superiors as to whether he could cope or, for that matter, whether he enjoyed the holiday. Scarcely suprisingly, after the first day back, his nervous symptoms worsened and he was then on sick leave for weeks. Reverberations from the 'Big Bang' in the City of London, from computerisation and from 'Black Monday', have caused psychological casualties amongst people who cannot cope with the increased pace and the rate of change. Having to cut corners goes against traditional training, and minor obsessional traits, which until then had stood people in good stead, can now be a hindrance, becoming more pronounced through stress, and let them down. Often mentioned, too, is the stress of interruption: of not being able to complete one task without having to switch to a completely different topic. To get anything finished then means taking work home.

As I have said elsewhere, repeated short-term sickness absence of one, two or three days may reflect job dissatisfaction. An excess of personally-significant stresses, be they work or domestic or both, causes not only psychological but also physical illness in a vulnerable personality. The analogy is the kettle full of boiling water that would explode without the means of escape for the steam up the spout.

Physical illnesses associated with stress are termed psychosomatic, and include some types of migraine, bowel disorders like irritable bowel syndrome, coronary artery disease and backache. There is also the possibility that people stressed by thwarted ambitions may be susceptible to some types of cancer; the same may apply to people who experience difficulty in forming relationships and expressing emotions.

To keep the subject of occupational stress in perspective, I like the comment of Harold Laski, the twentieth-century political scientist, who said 'Hard work is the cure for most ills', to which I would add, 'together with appreciation by others of that hard work'.

Counselling

The ready availability of managers to their staff is important: work has to be done, too, of course – so an ever-open door is not necessarily the answer. It is more a case of the manager letting it be known he or she is readily available, in a sympathetic, nonthreatening way, for confidential discussion about a problem, ideally on neutral ground away from the work environment. This may be time consuming but it is an important management function which can sometimes prevent a stress-related illness developing.

Some managers feel a non-work related problem is no concern of the company. This attitude may be strictly correct, but to expect anyone to be able to 'switch off' completely from any preoccupation, however serious, to stop thinking about domestic worries and concentrate solely on work, is surely unrealistic. Inevitably these problems intrude into the working day and have an effect: work efficiency falls, both of that person and perhaps their colleagues too.

Counselling involves discussion and devising choices, showing an interest and listening whilst the subject has the opportunity to explore problems and hopefully arrive at solutions – the counsellor does not provide the solutions. A long pause can be ended and the flow restarted by repeating the last phrase in an enquiring tone. 'I told him I couldn't take any more' becomes 'You told him you couldn't take any more?' To answer questions, give an opinion, or say what should be done, courts blame, should proffered advice, taken and acted on, not turn out as the person would wish; it is better to put a question back to the individual by asking 'and what do you think?' It is useful to know, or to know where to find, trusted sources of specialist opinion and voluntary help. If there is any possibility of illness, early referral to the company medical officer and/or GP is essential. A manager is not trained in and should not attempt formal psychotherapy – a mature and caring attitude is what is needed: taking an interest in the other person as a human being and perhaps revealing something of themselves to encourage the other person to self-disclose. To succeed in providing constructive help, any form of counselling must involve a psychological 'contract' between the helper and the would-be-helped. Both are often initially uneasy, fearful of disclosures and of the ability to understand and cope with the relationship.

Without the 'contract', discussions remain ill-defined, without room for manoeuvre; they readily become a charade, and then founder. Bad counselling does more harm than good and is worse than no counselling at all.

A psychological 'contract' is by no means the same as a commercial contract: it sets the scene for mutual trust. On non-threatening, neutral ground like a park, away from the work place, in a setting that will be quiet and undisturbed, the counsellor emphasises confidentiality and the need for joint, concrete problem-solving rather than the answering of questions.

Apparent empathy, in an atmosphere of genuine caring and understanding, must be created to allay any initial mistrust.

Usually several sessions are needed, and at each, progress is assessed and stock taken. The counsellor needs to be sensitive, unhurried and, mindful that a 'little learning is a dangerous thing', should preferably be trained; counselling skills are more than common sense.

Countering stress

In order to reach a top management appointment, an executive will have demonstrated ability to cope with work pressures and may be less in need of advice on how to relax and cope with stress than someone in a lesser position, either 'on the way up' or, particularly, if stuck and unlikely to achieve senior management. But, to avoid a stress-related illness, the work 'sponge' who takes on additional tasks whenever asked to do so must look critically at priorities, learn to delegate and to say 'no' when appropriate.

Methods of countering a stress reaction in an office setting after difficult calls or interviews include the following:

(1) Close your eyes and blank your mind by focusing on gentle, uninterrupted breathing, counting the breaths out up to ten.
(2) With interlocked palms gently flex the neck.
(3) Stretch one arm towards the floor and the other towards the ceiling.
(4) Take some exercise – go for a brisk walk during your lunch break.
(5) Stop all lunchtime drinking. Alcohol makes you drowsy and always slows you up. Keep work and alcohol separate.
(6) On occasions, stand up during a telephone call or meeting, or at least stretch in the chair.
(7) Imagine your body is a deep well. Close your eyes and each time you breathe out, drop a gold coin into the well through a small hole in the top of your skull. As you breathe out, in your mind's eye watch the coin wafting down through the well to waist level. Repeat up to ten coins – then start again.

Interviews about work for radio or television are stressful, particularly to someone unused to the experience. Stress can be countered by the following procedures:

(1) Discuss the whole programme with the producer – the length of the contribution and what it is intended to illustrate.
(2) Find out who else is to appear.
(3) If the programme is filmed rather than 'live', ask to see the extracts to be used.
(4) Have a rehearsal. Say what you want to say in short, succinct statements.

Alcohol is best avoided until afterwards. Such interviews are of course an extension of public speaking, in itself a stressful and unnatural activity, quite different from normal conversation, which is non-threatening and flows to and fro. The silence when the speaker stops is unnerving, so there is a tendency to hurry and avoid pauses. This is not the place to advise on the art of public speaking, but in essence what is needed is clarity, conciseness and the ability to convince.

Any speech should be well prepared, cover up to four main points, flow smoothly from one point to the next, and have an introduction and conclusion.

Nervousness is betrayed by fidgeting and staccato speech, so stand still, speak distinctly with pauses, and look at the audience. Above all, in any public-speaking situation, be natural and honest.

Staff training courses, like group selection procedures, can be stressful. This is particularly likely if there is an element of competition, or a need to prove ability, or if any kind of report on performance is to be submitted at the end of the course (or if members of the course mistakenly think there will be such a report).

A residential course will be more stressful than a day-time course, particularly to young people, some of whom will be away from home for the first time. Some courses deliberately aim to stretch mentally and challenge participants in a controlled setting under supervision; to do this is more realistic than providing a passive, sheltered, academic setting of learning by rote.

Any residential course, however intense or long the working day, must include time each day, usually in the late evening, when the course members are on their own, unobserved by tutors. Time for exercise or sport should also be included in the timetable, to use the pent-up energy produced by challenges on the course. This is never time wasted. Advice on how to make a presentation helps and course members must feel free to talk to tutors about their fears without loss of face. Tutors must be alert to the early signs of significant stress reactions. Not to be overlooked is the indirect effect of the course on the health and happiness of the employee's spouse. One business training college holds week-long courses to help the spouse understand the job and why it sometimes means time away from home.

Insomnia

This is an early symptom of stress – the physical stress of pain, or mental stress. The amount of sleep needed varies from person to person: the older you are the less you seem to need; between six and eight hours is average. Getting to sleep can be difficult for a few days after a flight which crossed many time zones, or after night-shift work. Persistent insomnia is a symptom of mental illness, especially depression – classically waking very early, feeling awful – but many healthy people fret if they temporarily have less than their usual hours of sleep. They ask for sleeping tablets but these can become a habit, especially as stopping a regular sleeping tablet causes disturbed sleep for a few nights.

Sleeping tablets are modified tranquillisers and though most are very short-acting, sedative and memory loss side effects could occur next day affecting concentration and performance. It should be established whether the employee who is regularly late for work in the morning, or who has become less effective, is taking a sleeping tablet at night and/or tranquillisers by day. Many people want tranquillisers, not to cover a crisis, but to help them cope with everyday life. In the sixties and seventies UK prescriptions for tranquillisers went up and up – nine million in 1964, sixteen million in 1970, over twenty million in 1975. Tranquillisers settle the symptoms of anxiety, but do nothing to remove its cause. Unless this is anyway temporary or can be readily resolved, there will be repeated requests to the GP for further supplies. For example, whilst tranquillisers calm the acute grief reaction to bereavement, this grief must be worked through. Tranquillisers delay and prolong this process. Tranquillisers depress brain function and take the edge off performance. Of course someone who is very anxious won't perform very well either. This is particularly relevant where their activities involve precision and skill, for example a salesman, or a chauffeur driving a company car. To have alcohol when taking a tranquilliser impairs brain function further. To take a sleeping pill plus alcohol on a plane risks transient global amnesia – coming-to hours later in the hotel room. Most tranquilisers are long-acting: a morning dose works all day.

In most situations, for most people, tranquillisers are safe and have predictable effects, but occasionally there is the reverse of the intended effect with intense excitement rather than calming, and aggressive behaviour due to the suppression of inhibitions. Taking a tranquilliser for a social reason, before a stressful interview for instance, is not recommended. The story is told of the student taking tranquillisers before a written examination, coming out two hours later very pleased with his efforts and then finding he had written his name over and over again. One advantage of tranquillisers over barbiturates is that they are much safer. An overdose of a barbiturate is life-threatening, whereas an overdose of a tranquilliser will likely cause no more than a prolonged sleep. Tranquillisers also act as muscle relaxants. In 1987 up to three million people in the United Kingdom were dependent on tranquillisers and would suffer significant withdrawal effects if the drug were stopped suddenly.

Caffeine

A common, seemingly harmless way that millions of people use to buck themselves up is to have a cup of tea or coffee. Indeed the commonest international stimulant is caffeine, which in one form or another is in tea, coffee and cola drinks, as well as in some pain-relieving tablets.

In most offices, tea and coffee are 'brewed-up' in the staff room or out of a vending machine; where they are freely available, daily intake can be very high. Totting up the daily intake can show a very high total which might have an effect on health. A cup of tea contains 60 mg of caffeine, a cola drink 50 mg, instant coffee 65 mg, and freshly brewed coffee 80 mg if percolated and 115 mgs if prepared by the drip method.

Alertness and the countering of drowsiness and fatigue are helpful effects, except at bedtime, but intake of more than 300 mg. of caffeine a day can produce a range of anxiety-like symptoms including restlessness, a fine hand tremor, palpitations and frequency of urination, plus rise in blood pressure and ringing in the ears. Going without caffeine for three to four hours can cause withdrawal effects of a headache and irritability in someone who has, inadvertently, become addicted to caffeine. Anyone in this state should cut down on caffeine gradually – drink fewer cups of these beverages, and weaker tea or coffee; try switching to decaffeinated products.

A person who is forever drinking coffee or tea may be in need of help, suffering perhaps from the first stages of a depressive illness and using caffeine to overcome symptoms of mental weariness, poor concentration and so on, in an attempt to keep going.

Anxiety

Anxiety is necessary for self-preservation. When walking along a quiet street at night, to be suddenly confronted by a man with a knife would make anyone anxious and produce rapid heart rate, sweating, and tense muscles, summoning one's all to fend off or run away from the threatening person. Without this anxiety reaction, the human race would be extinct.

The problem today, particularly in an office setting, is that the stresses are not caused by physically-threatening predators, but by overdemanding or hostile people, too great a volume of work, constant interruptions or impossible deadlines. These situations evoke the same anxiety reactions, which are not needed, cannot be used to advantage, and are a disadvantage, in that work must go on as if nothing had happened.

Overt anxiety in others, whatever the underlying cause, is not difficult to detect. Observe body language. That person is uneasy and restless, fidgeting – and if a nail biter, constantly nibbling. Sweating increases, particularly on the face, where it can be seen, and also from the palms, fingers and armpits. There is a fine tremor of the fingers and even of the whole hand, more apparent on movements which require accurate co-ordination. For example, picking up and drinking from a cup or glass. If anxiety worsens to agitation, the whole body may shake. The breathing pattern changes to chest breathing and the rate speeds up from the normal twelve to fifteen times a minute; this may cause faintness with tingling of fingers and feet. Sometimes the rate of breathing remains normal but the normally imperceptible rhythm changes to deep sighing.

An anxious person may complain of palpitations, the heart thudding as the rate increases from around seventy to well over a hundred a minute. There may be nausea and even actual vomiting; appetite goes and if the anxiety lasts more than a few days, body-weight falls. The person looks tense and worried, does not smile much or only fleetingly and is preoccupied. Obviously there is a loss of working efficiency. Apparent anxiety may have a physical cause – an overactive thyroid for example, and when consulted, the GP will exclude such conditions. Anxiety exacerbates the group of illnesses termed 'psychosomatic', where mental state appears to contribute to physical illness. Examples are varieties of asthma and skin and bowel disorders.

Anxiousness is a symptom of many nervous conditions besides 'anxiety state': obsessional state, hysterical reaction, and some varieties of depression, for example. Depression may follow prolonged unresolved anxiety. As we have seen, tranquillisers suppress anxiety symptoms but do not deal with the cause. This takes time. Strategies include supportive psychotherapy, use of biofeedback devices, learning to relax both mentally and physically in the stressful situation, hypnosis, behaviour therapy, positive visualisation, hand warming, deconditioning of maladaptive patterns of behaviour, and long-term psychoanalysis to find out why these responses occur. Also helpful is assertion training – learning the difference from aggression and how to react in an adult, mature manner when crises threaten.

ANOREXIA NERVOSA

This illness is an extreme way stress manifests itself. It usually strikes young women: the ratio is fifteen women to one man. The effects are physical but the cause is psychological, though still not clearly understood.

Typically the illness develops in mid-teens in a conscientious girl who develops a phobia about her weight, possibly sparked off by teasing or the imminence of adulthood, which she is trying to put off, and who has a false impression of her body image – she thinks she is bigger than she really is. Sometimes there are domestic tensions or overambitious parents, and maybe this is the one way she can get back at them. There is a compulsive slimming – food avoidance is ingeniously disguised and dramatic loss of weight follows. The girl claims to be eating normally and feeling well while doing all she can not to eat, surreptitiously concealing, sometimes in extraordinary places, food she says she has eaten, or making herself sick after meals or eating binges (bulimia), or both.

Periods cease and downy hair grows on the face. Whilst initially very active, denying anything is wrong, the girl sooner or later becomes lethargic and depressed; there is risk of suicide.

The illness is fatal in 5 per cent of cases. The GP who knows the family is well-suited to treat a mild case. Hospital in-patient care is indicated for the more severe cases, preferably in a specialised unit. Treatment is based on drugs, psychotherapy and behaviour therapy and can be successful, but takes time. To achieve initial weight gain may require intravenous feeding. Relapse can occur and long-term follow-up is necessary. Anorexic Aid is a society run by women who have recovered from this condition; it aims to support and advise sufferers of anorexia nervosa as well as family and friends. Each of the network of 500 voluntary self-help groups around the world is organised by a contact, and tries, in collaboration with medical and social services, to provide the insight and understanding essential to recovery.

HYSTERIA

'Hysteric' and 'neurotic' are undeservedly used as terms of abuse. Hysteria is a genuine condition in which, unlike malingering, there is no conscious motivation to deceive. It is not helped by slapping the face, and is diagnosed by exclusion and by recognition of subconscious motivation. Motivation consists of primary and secondary gain. The stressful problem is temporarily resolved by development of the hysterical symptoms. This is the primary gain. Secondary gain is the pleasure from the attention and sympathy gained from the symptom, which may be so enjoyable that the symptom continues when the original problem has been resolved.

The hysterical personality craves attention and affection, but is unreliable, superficial and casual towards others. Classically there is the tendency to overdramatise. More than one hysterical symptom may develop but this is rare; one is usual.

Symptoms include amnesia (memory loss termed 'hysterical fugue'), paralysis or loss of feeling in one or more limbs, blindness or deafness. Unless the hysteric has detailed medical knowledge, the symptom does not match a real medical condition, being the person's idea of how the condition would present. Also, despite what is an appalling catastrophe, sudden blindness for instance, the hysteric is not unduly distressed and is unnaturally calm – this is termed 'belle indifférence'. Treatment involves searching for stresses and analysis of social circumstances. Drugs and psychotherapy may be required. Hysteria is rarely diagnosed nowadays, so often in the end a physical cause is found for what was first thought to be hysteria.

OBSESSIONAL NEUROSIS

This is an exaggeration of tidiness, neatness and 'checking to be on the safe side'. Someone with an obsessional trait or actual obsessional neurosis is at increased risk of anxiety or depression; obsessional traits then become more evident. Examples are repetitive, compulsive acts like checking or hand-washing. The person knows these are unnecessary and pointless, but does not feel comfortable unless they are gone through, and becomes anxious if prevented from carrying them out. If the act is enjoyed, though guilt may follow, it is not an obsession. Another symptom is rumination, the endless turning over in the mind of one thought.

Of inflexible temperament, the obsessional adopts this behaviour as a defence against insecurity, doubt and indecision. At best a nuisance, little more than normal prudent checking of switches and locks, this neurosis can be totally disabling, with the whole day taken up by repeated rituals. Nothing is achieved and home life and job are jeopardised.

Seen in early adult life, the untreated condition persists for years, fluctuating in intensity, with exacerbations being linked to stress and depressive illnesses. In some cases there is a good response to psychiatric treatment and occasionally there is spontaneous clearance.

VIOLENCE

Along with the sharp, national increase in violent crime in recent years, it would appear that assaults at work are increasing: in the transport industry, for example. Attacks on doctors have also increased. Employers cannot always eliminate the risk of violence but must take reasonably practicable (the words of HASWA) measures to protect staff from attack. Should violence occur at work and an employee be injured, he or she may successfully claim damages from the employer (if able to prove negligence), as well as from the attacker. The injured employee may also apply to the Criminal Injuries Compensation Board for an award. Under the Social Security Act, victims of criminal attacks at work who are injured may claim industrial injury and/or sickness benefit. If death follows, their dependants may claim industrial death benefit.

In any work setting where money is handled as cash or there are other readily moveable valuables, there is the risk of attack. In a bank, the obvious example is a raid on a branch. When the raid is over and the police have left, staff involved in such a raid are likely to suffer a reaction of some kind, and the well-meaning tot of brandy or cup of tea, accompanied by words of comfort and advice to go home early, may not be enough.

A disabling psychological state may develop in a person who appeared totally calm at the time and who gets a delayed reaction. This is an example of where the traditional British 'stiff upper lip' is unhealthy; far better for avoiding long-term problems is to let the emotions out and, for example, to have a good cry.

Some people may go beyond this into a state of nervous shock with trembling of the limbs and over-breathing. For them a calm reassuring manner is therapeutic, whilst they get the tension and anxiety out of their systems.

Some people cannot believe a raid is really happening and feel it is all a nightmarish dream. They may feel numb, unable to move in a state of frozen immobility, which is a type of protective adrenaline fear reaction.

In the days and weeks after a raid, it is natural for staff to feel edgy and tense at work, with reminders all around of the disturbing drama of the actual raid. These feelings fade in time, but whilst they are happening, the following relaxation exercise practised from time to time will help.

Sit down comfortably in an upright chair in the rest room; legs uncrossed, hands on thighs; close the eyes and concentrate on breathing slowly, quietly and gently in through the nose and out through the mouth. Without pausing between breaths, and on each 'out' breath, think the word 'calm' and imagine tension flowing from you.

The most important and therapeutic role for any counsellor is to encourage the person who has been involved in a raid to talk about their experience, their dismay, anxiety and terror, and not to try and jolly the person along. It is important to have empathy and to show that you feel for the person and in no way to play down, make light of, or criticise them for what has happened. The person who has been involved must be encouraged to open up, rather than to bottle up. Adopting the same posture helps to show that the counsellor is interested and caring.

The person involved in a raid will, at the time, have felt very vulnerable and helpless, unable to do anything about it, and these feelings may persist and are similar to those experienced after a burglary at home.

The member of staff may not want to worry their family and so may not talk about the raid at home, but this is wrong as the person, though helped by counselling at the bank, also needs the loving care of those who are close, who can only help if they are aware of what has happened. Otherwise, there can be irritation and mistaken advice to pull yourself together and snap out of it, when what is really needed is patience and sensitivity.

A company nurse or doctor should visit next day to explain the normality of the psychological symptoms during and after a raid, and to encourage talking about it both collectively and individually. The person who lives alone is most vulnerable to a stress reaction. Individual counselling should also be made available and a follow-up visit made about six weeks after the first.

When a member of staff or the public has become aggressive and is violent or threatening violence, it is important to remember he or she may be ill and involvement of the company doctor or nurse may help. Where such assistance is not available, a calm, confident and understanding manner, which shows evident concern for the individual's problems, will often restore order and in a difficult and demanding situation, quiet reassurance should enable the person to be calmed down. If medical help has then arrived, treatment may be accepted.

To show annoyance or reciprocal hostility can be disastrous, with rapid escalation of the situation. Verbal retaliation may set off explosive rage.

Illnesses in which violent tendencies may occur include schizophrenia, mania and hysteria, and such physical conditions as a post-epilepsy state and hypoglycaemia (low blood sugar), either spontaneous (which is rare) or occurring in a diabetic on insulin, whose dose is too high or not balanced with his or her diet.

Alcohol or drugs, or a combination of both, are responsible for many episodes of violence.

Release of aggression is associated with drug addiction, to barbiturates especially, where the person can be passively demanding, abusive or violent. Amphetamines cause excitement and agitation which can turn to paranoid aggression. LSD causes a range of mental states from euphoria to terror. Ecstasy from cocaine, crack and crystal can turn sour.

Should violence occur and physical restraint become necessary to avoid further injury, restraint should be by sufficient numbers to be overwhelming and preferably by men, but with at least one female if a woman is to be restrained.

The minimum effective force should be used, but be prepared for unexpected strength as restraint is resisted, and keep out of range of shoes, teeth and nails, to avoid injury from kicking, biting and gouging.

The advice of the Psychiatric Social Worker (PSW) (formerly the mental welfare officer), is invaluable in situations where the person remains acutely disturbed. The PSW can be contacted through the local government offices and very often will attend to assess the situation first hand.

The PSW is fully conversant with the Mental Health Act. First passed in 1959, this Act repealed all the former relevant Acts. Of the more than 150 sections of the Act, the most relevant for management of the acutely disturbed is Section 29. This makes provision for emergency compulsory admission to a psychiatric hospital for observation for up to seventy-two hours. Application is made by the PSW, who has a duty to say if admission is considered desirable, or by a relative who should in any case be contacted if possible. The application requires the support of a doctor – not necessarily a psychiatrist – who judges that emergency detention is justified. Section 29 is used when a person needs to be detained, either for their own immediate protection or for the immediate safety of others, when delay would be dangerous. The PSW, with the help of the police, if necessary, will take an unwilling person to hospital; a sedative injection may first be given by a doctor.

After seventy-two hours, detention for a longer period will require further application under another section of the Act, usually Section 25, which authorises compulsory detention for up to twenty-eight days. The applicant completes the prescribed form and the application must be supported by two doctors, both of whom have examined the patient together, or within seven days of each other. One of the doctors should be a psychiatrist and the other the family doctor.

PHOBIAS

Living in a high-rise block of flats contributes in some people to stress-related illness. A study of wives in such accommodation suggested that the incidence of nervous illness increased progressively from the fourth floor upwards. However, their husbands reacted in the opposite way; the higher they lived, the better they felt. Unlike the wife, who was 'trapped' in a flat with young children, the husband went out to work each day, and, presumably, when at home enjoyed the view.

Similarly, employees in a high-rise building know that they are only there a limited time with the opportunity to have a break outside the building at lunchtime. There is no evidence that illness is caused by working in a high-rise building. Some people have a phobia for heights, feel uncomfortable when above a certain floor, which varies from person to person, and suffer anxiety symptoms, up to and including panic.

This is not a major problem in the United Kingdom, perhaps

because there are few skyscraper office blocks; also, staff with a phobia for heights who resign rather than move to such a building, may give another reason for leaving, not wanting to admit to what they consider a weakness.

Sometimes, in an interview, the true reason is given and then referral, through the GP for behaviour therapy can be curative. A few make no secret of their phobia; one person called for a disciplinary interview on the 20th floor startled the manager by saying 'I can't stand heights'. He then promptly lay down on the carpet. End of disciplinary interview!

Others have phobias about lifts, often developed some time after an episode of being trapped for several hours in a broken-down lift.

Some people feel giddy looking out of floor-to-ceiling picture windows in high-rise buildings and can be disturbed by the illusion that there is nothing to stop them falling out of the window to the ground outside; this can be helped by a 'no entry' zone alongside the windows, by painting a horizontal line on the glass at waist level or by sticking a horizontal line of tape on the window.

Many people have, and are able to live with, irrational fears; these are usually confined to one specific object, mice for example. Most people are able to organise their lives so that there is little risk of coming into contact with the feared object. In a minority, however, the irrational fear or phobia is overwhelming and dominates their lives, greatly limiting what they do. The origins of the phobia can sometimes be traced back to childhood when a parent had a similar fear; or to a specific psychologically traumatic trigger event. The obsessional personality is at most risk, due to a tendency to orderliness, the need to be able to predict and lead a well-established routine life – useful features to a certain extent but not when exaggerated.

In travel phobia, there is a fear of travelling by particular modes of transport, usually plane, train or tube, where there are periods spent shut in, unable to stop and get out. This phobia may develop weeks or months after an episode where the person has been in, say, a crowded train unexpectedly held up between stations by power failure.

Alternatively, the fear is not so much the particular mode of transport, as of being carried further and further away from the security of the home. In either case, the typical story is that the person who has resisted these fears, then gives in and either reaches the station but does not board the train, or gets on the train but leaves it at the next station and goes home.

A few people have a phobia of being pushed from a crowded platform into the path of an oncoming train or of falling from the platform onto the live rail.

'Anxious and depressed' on a phobic's medical certificate may be correct, but does not mention the underlying cause.

A typical development is for a phobic to say they will be in the next day and then not turn up. Then, where the firm has many branches, the request is made for transfer to an office near home. This course of action may seem the only way the company will have the benefit of that person, but once transferred, it is unlikely the phobia will be treated, since the phobic, preferring to remain in work near home, resists the discomfort of treatment.

A teenage girl had worked in the City since leaving school. She was offered a move to another office in the West End with better prospects. Her journey had been a crowded one by train, but now in addition, there was a journey on the underground, very congested at peak times. She began to feel sick at the thought of the journey, which now took twice as long.

Once she fainted on the train, and another time was so upset on reaching work that she had to be sent home by car. At weekends she felt better, but by Sunday evening she would be worrying about the prospect of another five days' travel. Her doctor gave her medicine to settle her stomach, but this did not really help. Finally, one day, she had to leave the train and go home. Her doctor recommended that she work locally, and she preferred this to behaviour therapy, in spite of the adverse effect a move out of London would have on her career.

When there are no local branches, resignation may be offered, sometimes for a fictitious reason, out of embarrassment to tell the truth; or a request is made to the GP to recommend early retirement on health grounds. Treatment involves facing the individual with the feared situation, either suddenly ('flooding') or gradually ('desensitisation'). Success is likely in an otherwise healthy person who has a single phobia. In 'flooding', the patient is taught to relax and stay in the phobic situation until anxiety is controlled (that is, they are thrown in at the deep end): 'Desensitisation' is kinder, and involves learning to relax fast when exposed to increasing levels of the stress over several weeks. Local attachment is helpful whilst arrangements are made and treatment is given.

Diagnosis depends on subjective evidence and could be used as a means to obtain local transfer by an artful employee who would just prefer to work near home and avoid commuting; many staff working in the centres of cities and living in the suburbs and beyond would welcome such a move, but rarely, if ever, manufacture such symptoms.

In agarophobia, there is disabling fear of open spaces and/or crowds, and inability to venture far from home. There may be other associated nervous conditions. Behaviour therapy is again often beneficial.

BEREAVEMENT

Life expectancy at birth is about seventy-two for men and seventy-six for women. Heart disease causes 40 per cent of all deaths, cancer 30 per cent, lung disease 15 per cent, stroke 12 per cent and accidents 3 per cent.

Each year in this country twelve in every thousand of the population die. This figure has not changed for thirty years, and totals around 650 000; 1800 deaths each day. 60 per cent of deaths are in the over-seventy age group. In the average GP's practice of 2500 patients, there are twenty-six deaths each year.

Occasionally, sudden death from 'natural' causes occurs at work, usually from a massive heart attack. A doctor is needed to certify death, and attempted first-aid, cardiac massage and the kiss of life should be continued until this happens or, more usually, until the ambulance crew arrives in response to a 999 call, and takes over. If there is any question of a death not being from natural causes, for instance, with suicide, the police are called. A company doctor on site can certify death and inform the local Coroner's Officer, who will arrange removal of the body.

If an employee dies at work, the manager or personnel officer is likely to be the one to break the news to the relatives. The manager is also involved in bereavement when granting compassionate leave, time off to attend a funeral, and so on. The grief reaction following bereavement, or in the time leading to the expected death of a loved one, is often profound. Medical help may be required, both in counselling and in providing the 'crutch' of sleeping tablets, tranquillisers or both.

Tranquillisers are best taken for only a short time to avoid dependence. Very helpful in acute distress, such drugs are long-acting. One dose has a 'half-life' of about three days, during which time the brain's efficiency is reduced, decision-making becomes more difficult, and so on. Grief must be experienced and sooner or later 'will out'. The grief reaction follows a pattern, one stage often merging into another, the intensity varying from person to person. Skilled help is needed at any stage if a person appears to get 'stuck'.

The initial, acute, profound, emotional response is followed by a period of unresponsiveness, an inappropriate, controlled calm lasting several days. Next there is a period of sadness and depression followed by the blaming of oneself and others.

The next stage is overemphasis of the dead person's virtues, then apparent adaptation to and acceptance of the loss. Finally, a balanced view is taken and the mourner re-emerges to normal life. Tranquillisers delay this process. Attempts to minimise grief can lead to a reaction of pathological intensity which leads to a depressive illness.

Delayed grief may take the form of considerable hostility and aggressiveness.

The social custom of leaving the bereaved alone 'out of consideration for their sorrowing', and the resulting social isolation, complicates the mourning process and can prolong it.

The state and most companies make provision for the wife should the male employee die during his working life, with a death-in-service grant and widow's pension. But there is no state widower's pension. Some companies have widower pension schemes.

Verification is required by the employer before agreed benefits are paid to the dependents. Sight of the death certificate provides this. A death certificate may be issued by a registered medical practitioner who was 'in attendance' during the deceased's last illness and who should have seen him or her within the fourteen days before death; a locum or partner, who only sees the body for the first time after death, may not issue a certificate.

If these conditions are not fulfilled, the Registrar of Births, Marriages and Deaths is likely to refer the death to the Coroner, who, after enquiry, will decide whether or not a *post mortem* is necessary. This causes little delay, but the phrasing used by the doctor on the causes-of-death section of a certificate, especially Section II, 'other significant conditions', may cause delay.

Where there are young children, a widower has innumerable difficulties. It is vital that he keeps mentally and physically well. If the wife had a job, besides possible raised expenditure, there is reduced income.

A widower is particularly hard hit if he has to work long, inconvenient and inflexible hours, especially if overtime is part of the job.

The widower who has young children and no relatives to help him must arrange and pay for child-minding as well as housekeeping. If the father is unable to manage, the children will be taken into care by the local authority.

There is a high remarriage rate for widowers, four times as high as that for widows, although the three million widows in the United Kingdom outnumber widowers by almost four to one.

Children may react with delinquency to the death of a parent. One helpful source of information is the National Council for One Parent Families, 225 Kentish Town Road, London NW5.

DEPRESSION

The following hypothetical case illustrates the development of depressive illness, the stress symptoms and behaviour.

Early on a Monday morning in December a forty-five year-old manager who had spent his working life with one company, and was ambitious to reach a senior post, jumped from a window, falling fifty feet to the ground below. He had not felt right for six months – he had been tense and ineffective, with difficulty in concentrating. His GP had prescribed drug treatment that made him feel drowsy by day and even less able to concentrate, but still 'wound-up' inside. He would sleep during his lunch breaks in the park, trying to clear his 'woolly' head, and in the early evenings whilst watching TV. He used to wake early at 5 a.m., unrefreshed, was eating poorly and lost weight. His wife could see his work worried him, and urged him to talk to his Personnel Officer, but with an eye to the future, he was reluctant to show any sign of 'weakness'. However, he finally had to seek help one Wednesday: when deputising for his boss, who was away on holiday for the week, he found he could not, through indecision, give an answer to a customer's straightforward question. He told the customer to return next day and telephoned Head Office to say he could not cope. A 'stand-in' was immediately despatched, and a personnel officer drove him home, advising him to see his doctor, but it was forty-eight hours before he could get a five-minute appointment with his GP. The doctor changed the treatment and said he would write a letter to refer him to the local psychiatrist, 'but it may be a month or so before he can see you'.

The Manager was in despair; what he needed, but was unable to ask for, was immediate help. He brooded over the weekend and then acted.

Fortunately, he was promptly picked up by a passing police patrol, taken to hospital and lived; he only had a broken leg and no one else was injured.

This was reactive depression resulting from personally significant stresses at home and at work in an obsessional personality. At work he had been moved a year before, at short notice, and with little explanation, from an office where he had been happy and effective to a smaller office where he had to work in the same room as the junior staff who were offhand, unco-operative, and ganged up against him when he had cause to tell one of them off. A meticulous person, he could not get on with his work with a background of trivial chatter. A visiting colleague asked how he put up with it; this he saw as criticism. At home, about the time of the change of work location, his elderly father moved in with them. He himself got on well with him, but his wife did not, and she increasingly saw her father-in-law in her husband; the once-stable marriage started to disintegrate. So there was strain and an 'atmosphere' at home and at work, and seemingly no escape.

The fracture healed and with psychiatric treatment and a different drug he started to improve. After six months, stabilised on long-term drug treatment and still under hospital supervision, he was well enough to try a temporary job at Head Office. With his self-confidence dented, however, he now wonders if he has it in him to succeed in the executive post to which he had aspired. At home, his father agreed to move into a residential home nearby. The manager also changed his GP.

One half of GP consultations involve psychological problems. Depression is the commonest mental illness, often caused by loss, be it of money, of children leaving home, through bereavement, or of position or 'face'. The word 'depression' has several meanings of varying significance:

(1) The normal passing emotion of unhappiness and feeling miserable. Such a temporary depressed or anxious mood that lasts hours or days can be a normal reaction to particular life events.

(2) 'Reactive' depression, a disproportionate reaction to personally significant stresses in a vulnerable person who has difficulty in coping; this may be accompanied by anxiety and lasts indefinitely.

(3) 'Endogenous' depression, a psychotic illness which either appears for no apparent reason and may recur, or which follows an illness such as influenza or occurs after childbirth or during the menopause.

The difference between (2) and (3) is not that clear-cut and often there is something of both in a depressive condition.

In a mild or early depressive illness, the conscientious, depressed person keeps going, works on and hides the fact that all is not well by staying late or taking work home. Lack of sleep, or more usually waking unrefreshed in the early hours when the depressed mood is at its worst, with the coming of day to face, are typical symptoms. He or she may drink a lot of coffee and tea for the caffeine boost. By evening the mood may anyway temporarily lift a little. A depressed person looks unwell and may be unkempt as there is a lack of interest in appearance. A depressive illness is suspected from signs such as change in behaviour and change in appetite and weight. There is disinterest in what was previously followed with enthusiasm, poor concentration, a constant weariness and an all-pervading gloomy outlook on both self and the world.

Libido is reduced. A depressed person is inefficient and slow, with poor memory and concentration. Self-confidence drains away. He or she does not participate in meetings and is unable to be assertive with either subordinates or superiors. The apathy and slowness require medical exclusion of a thyroid deficiency as well as differentiation from side-effects of drugs of addiction. There may be ineffective tension and anxiety with tremor, palpitations and perspiration: one or more of these may be complained of to the doctor, with the underlying masked depression not mentioned.

Depressed people are sometimes disturbed by delusions. These delusions (false beliefs) are often related to health, being hypochondriacal ideas of imaginary illnesses. There may also be hallucinations, hearing imaginary voices, or seeing things which are not there.

Depressed people tend to self-deprecation, unjustified feelings of unworthiness, failure and guilt. Despite a good work record, once on sick leave they may feel unable to continue with their present job and level of responsibility, considering themselves not worthy of their job grade or capable of the work they had been doing. They may offer to resign, or seek early retirement or demotion to a lower grade, giving up years of experience and prospects of promotion.

The self-assessment may have been temporarily correct when the illness started. When trying to carry on working, handicapped by a brain not functioning efficiently, everything is too much. This can be the explanation when pressure of work is blamed as the sole cause of a breakdown. As the illness develops, normal work takes longer and there is the impression of overload. Sometimes a change of job is the right decision: the work may have been too much in quantity or quality or both for too long. However, rarely is it solely work stress that causes the depressive illness. The guidance of a psychiatrist, who knows both sides of the employment story, is invaluable.

Long-term decisions by the employer about a depressed person's working future should, where possible, be deferred, just as no long-term decisions about their working future or anything else should be taken by the depressed person, who may later regret those decisions when thinking clearly and rationally again.

Once cured there should be a return to normal efficiency. In any case, the manager should not consider, or agree to, a request for demotion without first making enquiries about the person's health. To repeat, as it is an important point: when the employee is suffering from a mental illness, particularly depression, long-term decisions about his or her working future should, if possible, be delayed until recovery except where the employee and the medical advisers, with the full facts, consider that the working environment, conditions of work, or the work itself, was a significant factor in the breakdown. Then, in a large company, it may be possible to arrange a move within the company to provide a lesser degree of responsibility, with all that that implies – lower grade, less pay and so on, or a transfer to another office at the same grade because of a personality clash, or a change to less physical work, to less overtime, and so on. Sometimes a job change may only be needed for a short time during convalesence whilst self-confidence is restored.

Whilst effective antidepressants and tranquillising drugs have re-volutionised psychiatric treatment, some people think a pill is available, effective, and necessary for every unwanted mood state; what they tend to forget is that counselling may be more effective, and stressful life events are sometimes better coped with without recourse to medication. Depression can clear spontaneously without treatment if the cause is removed; but it may take a long time, up to and beyond two years. The drug treatment for depression was revolutionised by the discovery of specific antidepressant drugs which correct out-of-balance brain chemistry, and start to relieve the symptoms of a depressive illness within two weeks.

It is important that the patient knows that the initial side effects of the drug can be tiresome for the first week of treatment and might discourage him or her from taking the treatment as directed. A dry mouth, blurred vision, difficulty in starting urinating and drowsiness should improve or clear within a week. These effects are less pronounced when a low starting dose is built up to the optimal level over a few weeks. Compliance is also improved by infrequent doses; for example, evening and morning, or evening only, with no need for a mid-day dose. Persistent side effects can sometimes be made less obtrusive by taking most of the dose at bedtime, so that they occur during sleep and have cleared by morning. A drug causing mostly drowsiness as a side effect will help a person with a mixed anxiety and depression illness to get to sleep and stay asleep. There may be a 'hangover' effect next morning and subsequent poor time-keeping, with lateness getting to work.

One type of drug, a monoamine oxidase inhibitor (MAOI), is prescribed less because of the risk of marked transient hypertension as reaction to certain foods or drinks. A reminder card lists the items to avoid, such as cheese, yoghurt, yeast extracts, broad-bean pods and Chianti wine. Any anti-depressant drug is continued for several months after clinical cure and then gradually tailed off; to stop too soon risks relapse.

The prognosis for the time it will take to recover from mental illness and the likelihood of recurrence is more difficult to predict than for physical illnesses: much depends on the speed of the response to treatment and the maintenance of that improvement. As a generalisation, psychiatrists tend, in my experience, to an optimism which is not always borne out by events.

A single episode of depression/anxiety during and in the wake of significant stresses, may not recur; recurrence becomes less likely with increasing age, itself a maturing process. However, 'endogenous' depression may well recur and the interval periods of normal mental health become increasingly shorter, the more episodes of depression that occur. Long-term use of a drug called lithium carbonate may prevent excessive mood swing. Regular blood checks ensure the correct dosage. Lithium may also prevent the less common hypomanic episodes that occur with or without subsequent depression; 'manic depressive psychosis' is the term used when both phases of this 'bipolar' illness occur.

For reasons not fully understood, electro convulsive therapy (ECT) can be successful in treating depression where drugs have not been. It is indicated in some cases of severe depression which have resisted drug treatment; in some cases of recurrent depression, and where there is a risk of suicide.

ECT is an alarming term for a painless procedure given under general anaesthetic it can be remarkably effective when given twice weekly, usually as an in-patient procedure. Six or fewer treatments can cure where drugs have failed. Any disturbance of short-term memory clears within a few weeks. Applying the shock to just one side of the head, the non-dominant brain hemisphere, reduces the possibility of memory disturbance but may be less effective.

A memorandum from the Royal College of Psychiatrists on this subject says the memory impairment which follows ECT diminishes fairly rapidly with time, but may increase, transiently with the number of treatments.

Reassurance that the memory soon returns to normal, at the latest within a few weeks, is important. If a patient returns to work within this time, it is important that the employer knows about this and makes allowance for any memory disturbance. Work requiring a good memory should be minimised; if this is impossible, a discreet check is carried out and due allowance made for forgetfulness.

SUICIDE

Suicide is an ever-present risk in anyone severely depressed and it is a dangerous myth that those who threaten to kill themselves never do so. Most suicide victims have sought medical help within the previous week, not necessarily for the depression, and may have spoken of their intention to colleagues at work, to relatives or to their GPs. This threat must always be taken seriously.

Suicide does not necessarily happen when the person is most deeply depressed, when the inertia acts as a brake. The attempt may come as treatment starts to take effect, if the apathy lifts before the depressed mood does. 'Genuine' suicide is usually planned in detail. Time, place and method are chosen carefully. Usually a drug overdose is taken where there is no chance of discovery.

Quite different is the impulsive overdose to attract attention, the 'cry for help', or the borderline overdose taken to shock an erring partner into 'mending their ways'.

Personnel staff can be concerned when there is the need to counsel or discipline an unsatisfactory employee, when that person has a history of mental illness, especially if this includes attempted suicide, for fear of another attempt – which this time may be successful. 'What if he or she leaves my office and deliberately walks in front of a bus. I would not want that on my conscience', is a typical concerned statement. Threat of suicide must always be taken seriously. Provided the person to be seen is not on sick leave, it is reasonable for a personnel interview to be arranged. If such an interview is judged to be necessary during sick leave for mental illness, like severe depression, it is important to first clear this with the member of staff's doctors. It is prudent for the company doctor to tell the general practitioner and/or psychiatrist what the company intends to do and when.

On average in the United Kingdom, two people every hour try to kill themselves; every day, eleven succeed – 200 000 attempts and 4000 deaths a year.

Suicide is the third commonest cause of death under the age of twenty-five; 80 per cent in this age range are men – a 30 per cent increase in ten years from 1977 to 1987.

Many more women than men attempt suicide, especially so-called para-suicide, as a cry for help. This difference is perhaps because men bottle up their emotions more than women. Men use violent means to end their lives, such as hanging or shooting, whereas women usually take pills. About 60 per cent of suicides and attempted suicides are carried out under the influence of alcohol.

The Samaritans are counsellors who provide a 24-hour service over the telephone. A list of points alert them that the person at the other end of the telephone is thinking of suicide – and in need of immediate medical help:

Caller withdrawn, and cannot relate to you.
Has tried suicide before.
Definite suicide plans.
Has been 'tidying-up' personal affairs.
Sounds anxious as well as depressed.
Has some painful physical illness and long-lasting sleep disturbance.
Has feelings of uselessness and if retired or out of work cannot accept this.
Isolation, loneliness, uprooting.
Having to live with few human contacts.
Lack of a philosophy of life like a religious faith.
Financial worries.

A dangerous period is when the caller is starting to get better and now has enough energy to kill himself or herself.

A doctor consulted by a patient who appears depressed will normally ask whether they have had suicidal thoughts. This does not put the idea into their mind and most who have contemplated suicide are glad to be able to talk about how they feel. Those who have not considered suicide will say so.

HYPOMANIA

In (hypo)mania the difficulty is to convince the person they are ill and need help. Unaware of abnormal behaviour, the hypomanic feels on top of the world. The hypomanic is tireless and overactive, brimming over with plans and projects and 'flights of ideas', often grandiose and unfeasible. These grandiose schemes are planned and started, only soon to be abandoned for others. A normally reticent person becomes outgoing, outspoken and full of self-confidence. With little time for sleep, the person soon becomes exhausted, but not before possible financial ruin or committing the company to an unwise venture.

SCHIZOPHRENIA

Schizophrenia is the term used for a group of disabling mental disorders which largely affect young adults. These psychoses, that is, illnesses where there is loss of contact with reality, are characterised by a disorder of thinking; an inability to communicate clearly; strange behaviour; delusions (false beliefs), and hallucinations. The latter are often auditory, with the individual hearing voices which are usually externalisations of their own thoughts.

The cause is unknown. In the United Kingdom, once schizophrenia has been diagnosed, there can be difficulty finding work, as employers are reluctant to employ sufferers, although there is a good chance, following recovery from an acute onset of the illness, that the individual will remain well with no recurrence.

The stigma of a schizophrenic label is avoided or delayed in some cases by the use of a more general diagnosis, such as nervous breakdown, nervous debility or depression (where the patient is also depressed). Schizophrenia is diagnosed more readily in the United States than in Great Britain, criteria for diagnosis being different in the two countries. The illness may have an acute or gradual onset, and in a third of cases affects a person who has a schizoid personality. The incidence of schizophrenia is almost 1 per cent of the population (some 150 000 people, most of working age) and of schizoid personality, 3 per cent of the population. People with this personality are extremely shy and withdrawn, may be eccentric and suspicious and have difficulty in forming relationships.

Schizophrenia typically first occurs in a young adult in the late teens or early twenties. The first signs of the illness are little more than exaggeration of the schizoid traits. Work performance inexplicably deteriorates and there may be delusions of persecution. There is withdrawn behaviour, apathy, academic failure following early promise, evidence of eccentricity and difficulty with personal relationships. The onset may be masked by depression, anxiety or agitation. There is difficulty in thinking and functioning logically and, in some cases, a delusion that an outside agency has taken over control of behaviour. Paranoid ideas may occur, although these are more pronounced in late-onset schizophrenia, appearing in early middle age.

Disorganisation of thought processes, but not of memory, causes inability to separate rational from irrational thought. These disturbing thoughts are bewildering to family and friends, but the patient often appears unconcerned by them.

This is the picture of 'simple' schizophrenia starting in the late teens or early twenties. Two other varieties – 'catatonic' and 'hebephrenic' – also typically first present in a person in their twenties, but 'paranoid' schizophrenia does not usually show itself until the late thirties. Variations of the normal turmoil of adolescence can give a superficially similar impression, particularly where schizophrenia is of acute onset.

Though schizophrenia is a serious illness, the outlook is by no means always poor. Most people recover from the first attack; the acute onset variety, treated promptly, has the best prognosis. About 50 per cent of cases recover completely and permanently or are kept well with drugs. Of the other 50 per cent, half have recurrent attacks and half deteriorate in spite of treatment. No causative link with stress has been found, although stress can precipitate a recurrence.

ALCOHOLISM

Alcoholism, also called alcohol dependence, having a drink problem, and problem drinking, has preceding stages of excessive and then harmful drinking, that are included in the World Health Organisation's definition of alcoholism: 'Those excessive drinkers whose dependency has harmful results'.

Excessive drinking once meant more than a bottle of spirit a week. This is thirty units of alcohol, each the equivalent of a glass of wine or a half-pint of beer; now that figure has been revised downwards to between fourteen and twenty one units a week, including alcohol-free days (for women the lower end of the scale, as their bodies tolerate alcohol less well).

Physical, mental and social harm may result from alcohol. Illness, accidents and violence, whilst drinking, after drinking, or on withdrawal of alcohol may occur. Long-term there is damage to the brain and the rest of the nervous system, to the heart and the liver; in fact, virtually any part of the body may be damaged.

Alcohol may be taken as a calming agent, but it acts erratically, tolerance develops and more is needed to obtain the same result. Treatment is preceded by assessment and 'drying out'. Sobriety is essential before the reasons for drinking can be probed. 'Masked' depression may be revealed or may develop during the process of facing and working through the difficulties which led to alcohol dependence.

In the United Kingdom between 1 and 2 per cent of the population have a drink problem, so every company of any size is likely to have several people in this state. Alcohol consumption in the United Kingdom has doubled in the past thirty years, and this increase is directly reflected in the incidence of alcoholism and liver cirrhosis.

Many admissions to hospital are for illnesses linked to alcohol, and convictions for drunkenness have risen. Until recently, these increases had gone largely unnoticed by the public as attention was focused on drug abuse.

Alcoholism is prevalent and the incidence of liver cirrhosis high in France and Italy, where drinking patterns differ. Wine, which is drunk from an early age, is the main source of the problem.

When discussing alcoholism, to speak of 'she' is now as appropriate as 'he' because of the growing number of female alcoholics. Figures for female to male ratios have changed in recent years from 1:6 to 1:3.

From an employer's point of view, pointers to the possibility of alcoholism include:

(1) Deterioration in quality of job performance.
(2) Increase in short-term, one- or two-day, absences (STSA) classically including Mondays, when the after-effects of a weekend's drinking make it difficult to face work at the start of the week.
(3) Unusual reasons for STSA that do not 'ring true'.
(4) Increased incidence of accidents in the office or home; falls caused by unsteadiness and poor balance, following drinking for example.
(5) Increased incidence of road traffic accidents.
(6) Breath smells of alcohol early in the day – or constantly of a peppermint disguise.
(7) Drinking visibly more and faster, for instance, at an office party or official function: the alcoholic develops tolerance and so needs more alcohol to produce the same effects.

Regular intake of more than four units a day, that is, two pints of beer, four single tots of spirit or half a bottle of wine (4 glasses), may damage many parts of the body, especially the liver and nervous system. Fibrous enmeshing of liver cells (cirrhosis) blocks blood flow, causes back pressure and in time bleeding from varicose veins at the junction of the gullet and stomach (oesophageal varices). The brain will also be damaged – every hangover spells death of more irreplacable cells.

This damage occurs not only to the alcohol-dependent, who cannot start or get through the day without alcohol, but also to those heavy drinkers consuming the daily quantities given above, year in year out. Those who boast they 'hold their drink well' may later suffer liver failure from cirrhosis and be heard to say then 'if only I had realised this would happen'.

All important is the early indentification of the alcohol-dependent person and, in liaison with the GP and company doctor, the offer of treatment on sick leave and full pay. For this scheme to succeed there must be genuine commitment by the drinker. This is more likely if it is made clear their job is 'on the line' if they fail to co-operate and performance remains unsatisfactory. Provided the person follows medical advice to the letter they have:

(1) Retention of status.
(2) Retention of pension rights.
(3) A job to return to.

Whether cure in the full sense of the word is ever achieved is arguable, as many experts consider 'once an alcoholic always an alcoholic', or rather a potential alcoholic; if any alcohol is drunk in the future, alcohol dependence will soon return. A few experts believe that return to controlled social drinking is possible, but for the majority this is 'playing with fire'.

The National Council on Alcoholism published the text of a lucid talk given by Dr J. S. Madden, Consultant Psychiatrist, entitled 'Alcohol and Industry'. From the facts given in this address, it is seen that the typical alcoholic in employment is a forty year-old middle- or lower-middle-class man, married and from a stable background, who has not been promoted for several years, has lost his job at least once before because of drinking, takes a bottle to work, drinks before starting work and then drinks at times throughout the day.

Drinking may have contributed to an accident at work and caused considerable loss of time from work, especially on Mondays; the average absence each year is a staggering eighty-six days. Alcoholism is rarely given as the cause for certificated absence.

The features most commonly noticed in alcoholics include: leaving the work station temporarily; lunchtime drinking; red or bleary eyes; altered mood after lunch; impaired and uneven work performance; absenteeism for a day or half-day; more unusual explanations for absences; vociferous talking; prolonged lunch breaks; shaky hands; hangovers at work; shunning of associates; flushing of the face and an increase of minor illnesses.

Some companies prefer not to face the possibility of alcohol dependence or alcoholism in their staff, and claim it does not exist. Even to discuss the subject with senior managers may be resented as having personal implications or threatening their own alcohol supply at work, despite emphasis of the fact that there is no harm from moderate social drinking (although more UK companies are now adopting the policy of many American companies and becoming 'dry by day' – banning alcohol in working hours and forbidding alcohol on the premises, backed up by searches of personal property and even random urine tests).

When someone at work appears to be suffering from alcoholism, colleagues may out of mistaken loyalty attempt, successfully for a time, to disguise the fact, to cover up deficiences and help out with work left undone, as the alcoholic becomes inefficient. 'There's no point in talking to Mr X in the afternoons' is a typical remark.

In the short-term, the sick alcoholic, sick because alcoholism becomes a self-inflicted illness, gets by with this misguided help, but the longer the condition lasts, the more damage is done, part-reversible, part-permanent, and the more difficult it becomes to stop.

A company policy on alcoholism agreed by all levels of the work force should be widely disseminated. Concern that such a policy might indicate that the company has a particular drinking problem is unwarranted.

For treatment to be successful, motivation is all important. Ideally the alcoholic must *want* to stop drinking and get better, but knowing that job retention depends on co-operation and a successful outcome, is almost as good.

Early temporary relapses for a few days are not uncommon, and allowance should be made, but no employer can be expected to tolerate indefinitely repeated relapses, or persistent failure of treatment or, in the face of evidence to the contrary, refusal to accept there is a problem. The efficiency and at times the reputation of the organisation may suffer. In these circumstances, consideration will need to be given to termination of contract on the grounds of ill health, in liaison with the GP, specialist and company doctor.

At every stage, the alcoholic must be kept fully advised of the employer's attitude and intentions, and be given every opportunity to put their point of view.

The initial treatment of the alcoholic may be residential – away from all sources of alcohol – at a unit specialising in the condition. This is available under the NHS but facilities in some areas are limited. Specialist private patient clinics and hospitals provide a four to six week stay, followed by regular out-patient attendance.

After assessment and 'drying out' to rid the body of the breakdown products of alcohol, the mainstay of treatment is psychotherapy: group discussion sessions, when a trained psychotherapist is present, and informal discussions at other times. A short, sharp education and self-awareness course as an out-patient, lasting three weeks, may achieve as good if not better results, especially if coupled with the use of Antabuse, a drug which causes a violent reaction if alcohol is then taken.

Alcoholism is linked in some cases to an underlying psychological disorder. When dependence on alcohol relates to depressive illness, an underlying fear or phobia, a personality problem or difficulty with social interaction, then drugs or behaviour therapy or both are used to try to resolve the underlying cause and if successful, remove the risk of relapse once a normal life of sobriety is adopted on leaving the unit.

As well as physical and psychiatric effects, there are social effects. For example, alcoholism affects many people other than the drinker: spouse, children, parents and other relations at home, colleagues and employer at work and (if a driver), passengers and other road-users. Alcoholism and alcohol addiction are emotive terms. When talking to the person involved, the family or colleagues, it is better to use the phrase 'problem drinker', which also includes those heavy drinkers who are not yet addicted, but are harming their lives.

Alcoholics Anonymous (AA) is a nationwide organisation which, by personal contact and regular meetings, run and attended by recovered alcoholics, provides counsel and support. AA promote permanent abstinence: 'one drink heralds return to alcoholism'. A feature of AA meetings is recovered alcoholics talking of their experiences. 'Al Anon' is an offshoot of AA for the alcoholic's spouse and 'Alateen' for teenage children.

It has been suggested that employers should actively look for staff dependent on alcohol and encourage them to seek help, but a manager is not a doctor. It is better to stick to noticing and acting on deteriorating performance and, if you suspect problem drinking, to suggest referral for help.

As far as prevention is concerned, this should focus on the young and the need for any drinking to be responsible, getting over the message that heavy drinking signifies neither maturity nor virility.

The Health Education Authority got this point across neatly, with a poster that showed the male symbol, a drooping arrow surmounting a circle, with the caption 'There is one part alcohol is sure to reach': in other words, alcohol causes impotence.

The alcoholic is rarely drunk; alcoholism is distinct from acute drunkenness. Drunkenness at work may follow a lunchtime celebration where the one-hour break has become two or three. Typically those affected are young, have an empty stomach (no breakfast), and are celebrating a special event – birthday, engagement, Christmas, or something similar. Spirits are steadily consumed, sometimes pressed on them by friends. These same 'friends' consider subsequent drunkenness a huge joke.

But drunkenness is degrading, far from funny, and potentially lethal, due either to the poisonous effects of alcohol or, when unconsciousness supervenes, mismanagement by these friends; for instance, if the tongue obstructs the airway or vomit is inhaled, asphyxiation results.

Drunkenness from lunchtime drinking comes on during the afternoon, and typically only when sobering-up procedures have failed do the friends involve the first-aider, intending to leave him or her to cope and clean up the mess.

If unconsciousness occurs, this is a medical matter. Life-saving action may be needed and include admission to hospital. (This does not enhance the reputation of the company concerned.) But when the extent of the drunkenness amounts to slurred speech, unsteadiness and sickness, the 'friends' should cope with the drunk, cleaning up vomit and so on.

Also, if the first-aider is dealing with such a case she or he is not able to help another member of staff, however seriously ill they might be; a person cannot be in two places at once.

DRUG ADDICTION

Whilst some people become addicted to alcohol others succumb to drugs of addiction that cause emotional or physical dependence or both. Many drug addicts have been or become 'problem drinkers'. The term chemical dependency covers both states. The official figures for drug dependence in this country underestimate the size of the problem. They only include people dependent on drugs covered by the Dangerous Drugs Act, of whom only a proportion become registered with the Home Office. The trend is upwards.

Most drugs are given slang terms based on the trade name or the colours of the tablets: 'green and blacks', 'moggies' and 'mandys'.

Injections are often given in unhygienic surroundings, such as public toilets, using tap water rather than sterile water to dissolve the crushed tablet or capsule powder. Syringes and needles used more than once are unsterile and, passed from person to person, will transmit such illnesses as hepatitis and AIDS. Local infection and abscesses at the injection site are common – even loss of a limb may result. Injections are direct into a vein, usually at the elbow crease where scars mark past infected injection sites.

Treatment consists of phased rather than sudden withdrawal of the drug, controlling the symptoms which occur or substituting a less harmful drug as an intermediate step. Then follows mental and social rehabilitation. General pointers to drug dependence include:

(1) Self-neglect: the previously neat and tidy person becomes scruffy, dirty and smelly.

(2) Malnutrition, if away from the parental home. Money once spent on food is spent on drugs.
(3) Repeated short-term sickness absence.
(4) Loss of interest and pride in work.
(5) Social isolation.
(6) Breaking the law; for instance stealing to get money for drugs.
(7) Those taking drugs by injection have scars on the arms which they hide by always wearing long sleeves even in hot weather.

Specific dependences

Brief notes on specific dependences follow. Dependence on more than one group is not uncommon; then the presentation is blurred.

Morphine type. This includes three derivatives of opium – heroin, codeine and morphine. The number of heroin addicts in the twenty to thirty-four age group has been increasing each year.

Emotional and physical dependence develops rapidly and is followed by marked tolerance – ever larger doses are required for the same effect. A state of relaxed euphoria is accompanied by sleepiness, lethargy and poor concentration. Excitement may occur, as may sickness. Breathing is shallow, the skin itches and the pupils of the eyes become very small – 'pinpoint'. Once dependent, the search for the drug and how to pay for it, are the only concerns of the addict.

Sudden withdrawal is known as 'cold turkey'. After eight to sixteen hours there is miserable nervousness, restlessness and anxiety, followed by sweating, runny eyes and nose, muscular twitching and cramps, increased rate of breathing, raised temperature, dilated pupils, sickness and diarrhoea: Reaching a peak in seventy-two hours, the condition subsides in five to ten days.

Amphetamine type. Amphetamines, best known for their use in wartime to maintain vigilance, were also prescribed, until the risk of addiction was realised, to suppress the appetite of those who wanted to lose weight. Tolerance develops and there is severe emotional dependence but no physical dependence and so no characteristic withdrawal syndrome. Unless large doses are taken, the picture is of deceptively near-normal behaviour. Great cheerfulness and talkativeness is followed in an hour or so by quietness and depression. Behaviour can be reckless with larger doses; there is tension, anxiety, agitation and aggression, mimicking severe mental illness. Mental and physical activity is increased; self-confidence grows with un-awareness of inaccurate performance. Tasks are carried out at a faster rate and the person thinks performance has improved whereas in reality it has deteriorated.

Withdrawal of the drug exacerbates any existing mental condition and may precipitate severe depression. With suppression of appetite, weight is lost. Amphetamine is gaining popularity as, unlike cocaine, it can be manufactured in the United Kingdom.

Barbiturate type. Barbiturates are now seldom prescribed by doctors, with the exception of phenobarbitone for epilepsy. Marked emotional dependence and very severe physical dependence develop gradually.

Barbiturates are usually taken by mouth, but some addicts dissolve the powder in water and inject the solution into a vein – 'main-lining'; given in this way, the drug is very irritant, causing local ulcers and abscess formation.

Tolerance is less evident than with morphine drugs. Effects include confusion; swings of emotion, with sudden highs and lows; slurred speech; unsteady gait and reduced mental ability.

Withdrawal effects start within thirty-six hours, last fourteen days, and include anxiety, muscle-twitching, tremor of the hands (worse when co-ordination is required), weakness, dizziness, sickness and disturbed sleep. Abrupt withdrawal is dangerous and can kill.

As barbiturates are seldom prescribed these days, there has been a switch to the abuse of tranquillisers, which carry less risk of physical dependence but can have very disruptive effects on behaviour.

LSD. LSD is not addictive, but is a very potent hallucinogenic; a minute pinhead dose has a great effect on the brain. STP – serenity, tranquility and peace – is the intention, but there is a risk, especially to the person unused to the drug, of a severe panic reaction which, although the drug is never taken again, can recur weeks, months or even years later, when it is known as a 'flashback'. Restlessness with exhilaration or depression may progress to delusions or hallucinations and the inability to separate fact from fantasy.

Glue. There has been a craze among teenagers, reaching epidemic proportions in some parts of the country, for the sniffing of glues to inhale the volatile solvent. There is rapid intoxication and also the risk of asphyxia from sniffing with the head inside a plastic bag.

Cannabis consumption is widespread in this country. According to those who advocate legalising cannabis smoking, five million people in this country have smoked cannabis at least once. Cannabis has a distinctive, lingering, musty smell, once positively identified, never forgotten. Hence the use of police dogs to sniff out cannabis. Supplies come from Pakistan, Morocco and the Lebanon.

Cannabis did not at first appear to have physical ill effects, but now there is evidence that heavy and repeated use has a permanent detrimental effect on the brain and lungs and is a contributory factor in some road accidents. People under the influence of cannabis describe themselves as 'high'. There is talkativeness and illogical laughter. The eyes are red, there is a tendency to cough, the mouth and throat feel dry.

PCP is a veterinary tranquilliser that is abused by young people in America. It has unpredictable short-term effects and with continual use impairs memory and intellect. PCP, known as 'Angel Dust', is a powder that can be sniffed, smoked or swallowed or dissolved in water and injected. There results a zombie-like detachment, hallucinations and later feelings of depression, when violent paranoia may occur.

Cocaine is highly addictive. The taking of cocaine is widespread in America and has spread to Britain. In America there has been much publicity about TV and film personalities who have disclosed a past history of addiction. In the United Kingdom the biggest users at first were apparently office workers; so much so that in some offices the men's lavatory, where drug taking went on, was referred to as the powder room.

The stimulant, 'high' effect of cocaine is similar to that of amphetamine but is more intense, comes on faster, is short-lived and leaves profound depression in its wake. Cocaine is sniffed ('snorted'), or injected, or 'cooked' with water and baking soda and then smoked. This latter process removes impurities and leaves a crystallised cocaine base called 'crack'. 'Crack' is cheaper than cocaine, acts in seconds and is just as quickly followed by a devastating, crushing 'down' of despair and depression which is only relieved by another 'fix'. Thus, 'crack' is instantly addictive from the first dose and so to even try it 'just the once' is dangerous and foolhardy.

A person 'high' on cocaine appears rather drunk but does not smell of alcohol. He or she is fatuous and hyperactive, has a flushed face and, if you look closely, dilated pupils. Someone who keeps taking cocaine will develop a range of physical and psychological symptoms. The physical symptoms can be the cause of sickness absence and include nose-bleeds, persistent runny nose and nose damage if the drug is 'snorted', weight loss, tummy upsets, chronic ill health and heart rhythm changes. Mentally, he or she is irritable, unreliable, suspicious (paranoid even) and dishonest. Nightmares and hallucinations of bugs crawling all over the body are other unpleasant effects.

Everyone should be alert to drug and alcohol abuse and aware of sources of help.

Summary of information on drugs

Most people do not take illicit drugs or abuse prescribed drugs, but of those who do, many start taking a drug out of curiosity and/or because they are offered the drug by someone they know. Other reasons are boredom, resentment, or to help as a coping aid (and this is the case with tranquillisers). Often a single experiment or two with drugs is enough, but some people go on and become addicted. It is far from easy to spot occasional drug-taking, but regular drug-taking, apart from the specific changes listed below, may cause the following:

Behaving out of character; changing from a bright, cheerful person to being moody and disgruntled, irritable and aggressive, subject to spells of drowsiness, losing interest in work, sport and hobbies, taking less care over appearance; money may go missing. Some drugs are taken by injection, when there is the risk that an unsterile needle may be used and cause infection, local abscesses which heal with scars; using other people's needles, contaminated with their blood, carries the risk of Type B hepatitis and of acquiring the AIDS virus.

There are many other explanations for the symptoms and signs listed in Table 9 but these may provide clues.

Treatment of drug addiction requires skill and perseverance. Doctors are required to notify the Chief Medical Officer at the Home Office if they suspect any drug dependence, or treat a heroin addict.

Education into the long-term consequences of continued drug-taking is an important part of treatment. There is at present a shortage of the residential care that may be necessary. Special Youth Advisory Centres are available in some areas.

Table 9 Information on drugs

Name of drug	Source and route taken	Early effects	Late effects	Type of dependence
Amphetamine	Synthetic powder. Oral, smoke, sniff. Inject.	Effects last 3–4 hours. At first lively and energetic; can get anxious and irritable, restless and panicky. Increased heart and breathing rate.	Delusions of persecution; depression; attempted suicide.	Psychological dependence.
Cocaine, a stimulant like amphetamine; also called coke and crack.	White crystal powder made from coca shrub leaves. Sniff (snort) smoke, inject.	Short lasting burst of euphoria and alertness; anxiety and panic may follow; then fatigue and depression.	Irritability, paranoia, hallucinations; bronchitis and lung cancer; damage to the nose from sniffing; damage to veins from injections.	Psychological dependence.
Cannabis (sativa); also called hash, dope, marijuana	Dried herb or sticky lump with characteristic clinging odour. Smoked as a cigarette with or without tobacco. Often in a group, or added to food or drinks.	Effects last 1–5 hours. Symptoms of intoxication but no smell of alcohol. Sense of relaxation, with talkativeness and great amusement.	Bronchitis and lung cancer.	Psychological dependence.

Drug	Form / use	Effects	Health risks	Dependence
Heroin also called smack	White or brown speckled powder from Opium Poppy; oral, sniff, smoke, inject.	Heightened awareness of colours and sounds. Some difficulty with speaking and writing may follow, with impairment of all skills and sleepiness. There can be red eyes and there is a high accident risk. Depressant; slows reactions; detached and relaxed; red bloodshot eyes;	Constipation; periods stop; damage to nose if sniffed; damage to veins if injected. Withdrawal effects like influenza. Heroin can be diluted by traders with other white powders (like flour and talcum powder) which increases risks.	Strong psychological dependence; also physical dependence.

334

Table 9 *continued*

LSD (Lysergic Acid Diethylamide)	Synthetic hallucinogen. Also hallucinogenic mushrooms.	LSD taken orally as tiny tablets or absorbed on sugar lump or piece of paper. Acts for up to 12 hours.	Experiences vary but include intense/disturbed emotions and senses. A bad trip causes depression, disorientation and panic.	Repeat experiences 'flash backs' can occur months later.
Barbiturates	Powder in capsules; oral with alcohol or inject. Effect lasts from 3–12 hours	Depressant similar to alcohol. Sense of relaxation; 'drunken' state; unconsciousness.	Withdrawal symptoms of restlessness, confusion and fits.	Psychological dependence. Physical dependence. Prescription only medicine.
Tranquillisers; Benzodiazepines e.g. Valium, Ativan.	Oral, tablets or capsules. Effect lasts 3–6 hours.	Helps anxiety, excess pressure, and sleep.	Tolerance may develop withdrawal effects include insomnia, confusion and anxiety.	Psychological dependence and physical dependence. Prescription only medicine.
Solvents	Vapours from glues, etc., inhaled	Symptoms of intoxication but no smell of alcohol.	Fatigue; poor concentration. Risk of damage to liver, etc.	

COMPULSIVE GAMBLING

Another addiction, one that causes social, financial and psychological difficulties, is compulsive gambling. To most people a pastime, for a few gambling takes over their lives. Betting shops and fruit machines provide the gambler with ready opportunity.

The Royal College of Psychiatrists, which submitted a paper to the Royal Commission on gambling, considers that too little is known about why and how much people gamble and about the distress it causes others. Its paper divides pathological gambling into five categories:

(1) Compulsive gambling – there is loss of control, gambling is irresistible and the gambler is unable to stop until all the money is gone.
(2) 'Sub-cultural' gambling, where everyone else in the environment gambles heavily.
(3) Gambling associated with a depressive illness. When this is cured, the compulsion to gamble disappears.
(4) Neurotic gambling – a reaction to an emotional problem or stressful situation.
(5) Gambling by a psychopathic personality – a person without morals who acts on impulse and does not learn from experience.

For all gamblers, in whichever category, there is 'Gamblers Anonymous', with branches nationwide which provides support for both gamblers and their families. The compulsive gambler when gambling is in a dream world and, like the alcoholic, is secretive about the compulsion and is unlikely to seek help until at 'rock-bottom'. Essential for successful treatment is admission of the need for it.

EMPLOYEE ASSISTANCE PROGRAMMES (EAPs)

These are well established in American industry and are starting in the United Kingdom. The principle is that any member of staff with any personal problem which is seriously worrying him or her is free to contact, at any hour of the day or night, an experienced counsellor who is not part of the personnel management of the company, with a view to seeing that person for stress counselling with no feedback to the company. The problem will be causing anxiety and whilst it is not necessarily directly work-related (and the experience is that the highest proportion of problems are marital), it will inevitably lead to loss of work efficiency.

The counsellor can be from a team of independent consultants who provide a 24 hour on-call service for an annual fee based on the total number of employees, or from a team of specially-trained staff from the company with professional counsellor support – an advance on peer co-counselling, which is something many employees do instinctively when they have a work problem, that is, talk about it to a colleague doing the same work. Company nurses and doctors can and do fulfil the role of confidential counsellor, in either of these latter two ways; an EAP can thus be 'in-house'. It is important for such people to be trained, as there is more to counselling than being 'good with people'. The golden rule is to guide, but not to positively advise – rather to let the distressed person vent his or her feelings and 'work out their own salvation', providing support whilst this happens, and information.

The problems range from marriage breakdown to financial setbacks and from alcohol abuse to inability to cope with some aspect of the work. Domestic worries cannot be switched off at the office door, will preoccupy people and so will affect their efficiency. Employers in America find enormous benefit to the company by way of improved productivity and reduced absenteeism and wastage with great costsaving. To succeed, it is essential that an EAP is separate, and seen to be separate, from management and the personnel department; staff must know they can speak in complete confidence.

Appendix: Sources of Help

ACCEPT NATIONAL SERVICES
Addiction Community Centres for Education, Prevention, Treatment and Research
Accept Clinic,
200 Seagrave Road,
LONDON SW6 1RQ.

ACTION ON SMOKING AND HEALTH
(ASH)
5–11 Mortimer Street,
LONDON W1N 7RH.

AL–ANON FAMILY GROUPS
61 Great Dover Street,
LONDON SE1 4YF.
(for the relatives of problem drinkers)

ALCOHOL CONCERN
305 Gray's Inn Road,
LONDON WC1X 8QF.

ALCOHOLICS ANONYMOUS
PO Box 1,
Stonebow House,
Stonebow,
YORK YO1 2NJ.

ANOREXIC AID
The Priory Centre,
11 Priory Road,
HIGH WYCOMBE,
Bucks HP13 6SL.

ARTHRITIS CARE
6 Grosvenor Crescent,
LONDON SW1X 7ER.

BACK PAIN ASSOCIATION
31–33 Park Road,
TEDDINGTON, Middlesex.
TW11 OAB.

BODY POSITIVE
LONDON 51 Philbeach Gardens
SW5 9EB.
(Help for HIV positive people by HIV positive people)

BRITISH ASSOCIATION OF CANCER UNITED PATIENTS
(BACUP)
121/123 Charterhouse Street,
LONDON EC1M 6AA.

BRITISH DIABETIC ASSOCIATION
10 Queen Anne Street,
LONDON W1M OBD.

BRITISH DYSLEXIA ASSOCIATION
Church Lane,
Peppard,
Oxfordshire, RG9 5JN.

BRITISH EPILEPSY ASSOCIATION
Ansley House,
40 Hanover Square,
LEEDS LS3 1BE.

BRITISH RED CROSS SOCIETY
9 Grosvenor Crescent,
LONDON SW1X 7EJ.

BRITISH TINNITUS ASSOCIATION
c/o 105 Gower Street,
LONDON WC1E 6AH.

CHEST, HEART AND STROKE ASSOCIATION
Tavistock House,
Tavistock Square,
LONDON WC1H 9JE;

65 North Castle Street,
EDINBURGH EH2 3LT;

21 Dublin Road,
BELFAST BT2 7FT.

COLOSTOMY WELFARE GROUP
38/39 Eccleston Square,
LONDON SW1V 1PB.

CORONARY PREVENTION GROUP
60 Great Ormond Street,
LONDON WC1N 3HR.

CRUSE, THE NATIONAL ORGANISATION FOR THE WIDOWED
AND THEIR CHILDREN
Cruse House,
126 Sheen Road,
RICHMOND, Surrey.
TW9 1UR.

DISABLED LIVING FOUNDATION
380–384 Harrow Road,
LONDON W9 2HU.

DRINKWATCHERS
Joe Ruzek,
200 Seagrove Road,
LONDON SW6 1RQ.
(Promotes safe and sensible drinking)

FAMILIES ANONYMOUS
88 Caledonian Road,
LONDON N1 9DN.
(A self-help support group for the familes and friends of drug abusers)

HAEMOPHILIA SOCIETY
123 Westminster Bridge Road,
LONDON SE1 7HR.

HYSTERECTOMY SUPPORT GROUP
11 Henryson Road,
LONDON SE4 1HL.

ILEOSTOMY ASSOCIATION OF GREAT BRITAIN AND IRELAND
Amblehurst House,
Chobham,
WOKING, Surrey.
GU24 8PZ.

INTERNATIONAL GLAUCOMA ASSOCIATION
King's College Hospital,
Denmark Hill,
LONDON SE5 9RS.

JOHN GROOM'S ASSOCIATION FOR THE DISABLED
10 Gloucester Drive,
Finsbury Park,
LONDON N4 2PL.

LEUKAEMIA CARE SOCIETY
PO Box 82,
EXETER, EX2 5DP.

LONDON CENTRE FOR PSYCHOTHERAPY
19 Fitzjohns Avenue,
LONDON NW3 5JY.

MANIC DEPRESSION FELLOWSHIP
c/o Council for Voluntary Service,
51 Sheen Road,
RICHMOND, Surrey, TW9 1YO.

MASTECTOMY ASSOCIATION
26 Harrison Street,
off Gray's Inn Road,
Kings Cross, LONDON WC1H 8JG.

THE MEDICAL COUNCIL OF ALCOHOLISM
1 St. Andrew's Place,
LONDON NW1 4LB.

MENTAL HEALTH FOUNDATION
8 Hallam Street,
LONDON W1N 6HD.

MIND (NATIONAL ASSOCIATION FOR MENTAL HEALTH)
22 Harley Street,
LONDON W1N 2ED.

MULTIPLE SCLEROSIS SOCIETY
25 Effie Road,
LONDON SW6 1EE.

MYALGIC ENCEPHALOMYELITIS ASSOCIATION
P.O. Box 8,
Stanford Le Hope,
Essex, SS17 8EX.

NARCOTICS ANONYMOUS
PO Box 246,
LONDON SW10 ODP.

NATIONAL CHILDMINDING ASSOCIATION
8 Masons Hill,
BROMLEY,
Kent, BR2 9EY.

NATIONAL ECZEMA SOCIETY
Tavistock House North,
Tavistock Square,
LONDON WC1H 9SR.

NATIONAL FEDERATION OF KIDNEY PATIENTS'
ASSOCIATIONS
c/o Mrs. Margaret Jackson,
Acorn Lodge,
Woodsetts,
WORKSOP, Notts. S81 8AT.

NATIONAL LEAGUE OF THE BLIND AND DISABLED
8 Tenterden Road,
Tottenham, LONDON N17 8BE.

NATIONAL SCHIZOPHRENIA FELLOWSHIP
78/79 Victoria Road,
SURBITON, Surrey, KT6 4NS.

ONE PARENT FAMILES
255 Kentish Town Road,
LONDON NW5 2LX.

PARKINSON'S DISEASE SOCIETY
36 Portland Place,
LONDON W1N 3DG.

PHOBICS SOCIETY
4 Cheltenham Road,
CHORLTON-CUM-HARDY,
Manchester M21 1QN.

THE PRE-RETIREMENT ASSOCIATION
19 Undine Street,
Tooting,
LONDON SW17 8PP.

THE PSYCHOTHERAPY CENTRE
1 Wythburn Place,
LONDON W1H 5WL.

THE ROYAL ASSOCIATION FOR DISABILITY AND
REHABILITATION
25 Mortimer Street,
LONDON W1N 8AB.

ROYAL NATIONAL INSTITUTE FOR THE BLIND
224 Great Portland Street,
LONDON W1N 6AA.

ROYAL NATIONAL INSTITUTE FOR THE DEAF
105 Gower Street,
LONDON WC1E 6AH.

THE SAMARITANS
17 Uxbridge Road,
SLOUGH,
Bucks,
SL1 1SN.
(Helping people who are feeling overwhelmed by stress and personal
problems)

SCHIZOPHRENIA ASSOCIATION OF GREAT BRITAIN
International Schizophrenia Centre,
Bryn Hyfred,
The Crescent,
BANGOR, Gwynedd, LL57 2AG.

SCOTTISH COUNCIL ON DISABILITY
Princes House,
5 Shandwick Place,
EDINBURGH EH2 4RG.

SCOTTISH MARRIAGE GUIDANCE COUNCIL
26 Frederick Street,
EDINBURGH EH2 2JR.

SCOTTISH SPINAL CORD INJURY ASSOCIATION
Princess House,
5 Shandwick Place,
EDINBURGH EH2 4RG.

SPINAL INJURIES ASSOCIATION
Yeoman House,
76 St. James's Lane,
LONDON N10 3DF.

THE TERRENCE HIGGINS TRUST
LONDON WC1X 8JU
(Trained volunteers offer help and support to people who are
HIV-antibody positive and those with AIDS, their friends and families)

TURNING POINT
4th Floor,
CAP House,
9/12 Long Lane,
LONDON EC1A 9HA.
(in the field of drug and alcohol misuse. Offering rehabilitation and care
to those with a drug or alcohol-related problem and support to families
and friends).

WALES COUNCIL FOR DISABLED
Caerbragdy Industrial Estate,
Bedwa Road,
CAERPHILLY, Mid-Glamorgan,
CF8 3SL.

WOMEN'S NATIONAL CANCER CONTROL
CAMPAIGN
1 South Audley Street,
LONDON W1Y 5DQ.
(Promotes early detection of pre-cancer of the cervix and cancer of the breast; mobile clinics)

Index